3 2044 049 131 675

W9-CGO-012

BUSHMANDERS AND BULLWINKLES

Mark Monmonier

HOW
POLITICIANS
MANIPULATE
ELECTRONIC
MAPS AND
CENSUS DATA
TO WIN
ELECTIONS

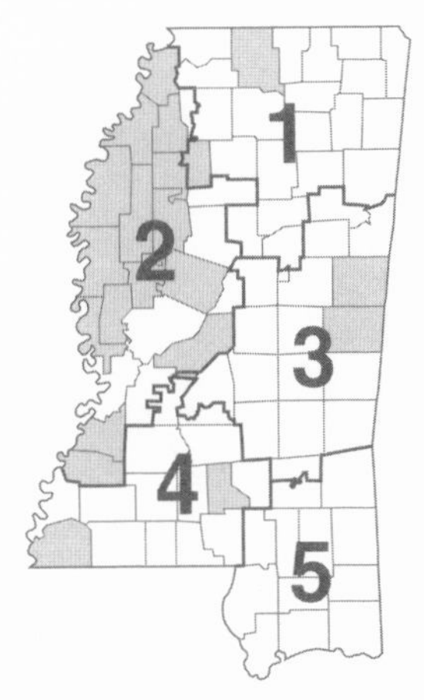

Bushmanders & Bullwinkles

THE UNIVERSITY OF CHICAGO PRESS
CHICAGO AND LONDON

A professor of geography in the Maxwell School of Citizenship and Public Affairs at Syracuse University, Mark Monmonier is the author of eleven books, including *How to Lie with Maps, Cartographies of Danger: Mapping Hazards in America*, and *Air Apparent: How Meteorologists Learned to Map, Predict, and Dramatize Weather*, all published by the University of Chicago Press.

The University of Chicago Press, Chicago 60637
The University of Chicago Press, Ltd., London

Printed in the United States of America
10 09 08 07 06 05 04 03 02 01   1 2 3 4 5
ISBN: 0-226-53424-3 (cloth)

Library of Congress Cataloging-in-Publication Data

Monmonier, Mark S.
Bushmanders and bullwinkles: how politicians manipulate electronic maps and census data to win elections / Mark Monmonier.
p.cm.
Includes bibliographical references and index.
ISBN 0-226-53424-3 (alk. paper)
1. Gerrymandering—United States. 2. Apportionment (Election law) United States. 3. Election districts—United States. 4. Cartography—Political aspects—United States. 5. Geographic information systems—Political aspects—United States. I. Title
JK1341.M66 2001
328.73'073455—dc21

00-060727

♾ The paper used in this publication meets the minimum requirements of the American National Standard for Information Sciences—Permanence of Paper for Printed Library Materials, ANSI Z39.48-1992.

For Sam Natoli, geographer, educator, editor, and advocate

CONTENTS

PREFACE

"Remap" is not in the dictionary, but it should be, as both verb and noun. Every ten years America counts heads, reallocates seats in the House of Representatives, and raises the blood pressure of elected officials and wannabe lawmakers by remapping election districts for Congress and state legislatures. And many jurisdictions also reconfigure city councils, town boards, or school districts. Because the way political cartographers relocate district boundaries affects who runs as well as who wins, a remap can strongly influence, if not determine, what a government does or doesn't do, what activities it bans or encourages, and which citizens absorb the costs or reap the benefits. Although "redistricting" refers to the process of drawing lines while "reapportionment" more narrowly connotes the reallocation of House seats among the states, neither term adequately alludes to the map itself as an object of debate and manipulation, if not litigation and ridicule.

Equally important is "gerrymander," a dictionary term with two shades of meaning. For many political scientists, gerrymandering is nothing more than deliberately increasing the number of districts in which a particular party or group is in the majority. They see this kind of manipulation as neither unfair nor illegal: if a state's constitution lets the party in power enhance its control, so be it. But for the media, the general public, and some judges, the term suggests sinister shapes that signify unfair if not illegal manipulation and undermine confidence in our electoral system. Although legal scholars differ on whether irregularly shaped districts are "expressive harms" that warrant judicial intervention, the post-1990 remap made unprecedented use of electronic cartography to craft cleverly contorted, racially motivated gerrymanders.

I wrote *Bushmanders and Bullwinkles* to promote an informed appreciation of redistricting, including the variety of plausible remaps and their diverse effects. In showing how boundaries can serve or disadvantage political parties, incumbents, and racial or ethnic groups, I want to make readers aware of how legislators, judges, and other elected officials use the decennial remap to promote personal or ideological agendas. Although the discussion frequently raises questions of fairness, my goal is not a cynical deconstruction of a fundamental political process but a skeptical, and I hope insightful, look at the complex relationships among geography, demography, and power. Despite my focus on congressional districts, the principles and ploys examined here apply as well to legislative districts at the state and local levels.

Although librarians and booksellers might shelve this book with works on political science, I approached the project as a study in the history of cartography in the late twentieth century—an examination of how legislators, redistricting officials, and constitutional lawyers use maps as both tools and weapons. In this vein, I emphasize two themes: the persuasiveness of cartographic propaganda and the technological momentum of computerized mapping. Detailed geographic databases and efficient interactive mapping systems have helped legislators and judges craft extraordinarily intricate districts designed to favor minority candidates, entrenched incumbents, or a particular party, and the Rorschach-like silhouette of the post-1990 remap has replaced the classic gerrymander cartoon as the emblem of perverse political cartography.

I focus on the map as a tool for both computation and communication. Within this framework, I examine the history of American political cartography, including relevant legislation, judicial decisions, and legal opinion. Chapter 1 introduces the book by explaining bushmanders and demonstrating why election districts matter, and chapter 2 provides the historical background for understanding legislative reapportionment in late twentieth-century America, including the law's vague preference for compact districts and our government's belated efforts to address discrimination against city dwellers and racial minorities. Chapter 3 pursues these themes through the 1990s, when the Justice Department and the Supreme Court sent mixed, often ambiguous messages with ominous implications for the post-2000 remap, and chapter 4 looks closely at the political impact of district boundaries drawn in the early 1990s. Chapters 5 and 6 offer a critical appraisal of traditional notions of compactness and contiguity as well as the role of maps in public hearings, news stories, lawsuits, and judicial opinions. Chapter 7 focuses on the prime goal of most redistricting efforts: protecting the seats of incumbents. Addressing new technology, chapter 8 looks at how geographic information systems have greatly changed the nature of redistricting. Turning to the data and racial categories used in redistricting, chapter 9 illuminates controversies over the 2000 census, which will merge statistical adjustment with traditional census enumeration and let citizens list more than one race. Chapter 10 finds attractive representational alternatives in early twentieth-century Ohio, contemporary Cambridge, Massachusetts, and much of modern Europe—alternatives that challenge our cavalier dismissal of cumulative voting, multimember districts, and other strategies that offer fair and effective representation without contentious cartographic manipulation. The epilogue recapitulates the vexatious issues of race, shape, and geography and explores challenges of redistricting during the first decade of the twenty-first century. In addition to forcing redistricting officials to confront two sets of census results as well as to accommodate growing numbers of mixed-race Americans, the post-2000 remap will test the restraint of politicians tempted to binge on highly contorted yet arguably constitutional nonracial gerrymanders.

Over fifty maps explain how various gerrymanders work and why some redistricting plans attract judicial scrutiny and public scorn.

Addressed directly in all cases, these maps are a vital part of the story and merit more than a cursory glance. Numbers tell another part of the story: counts of residents, minority voters, or registered Republicans (or Democrats) are, after all, what the maps represent and what their authors seek to either equalize or maximize. Although numbers are unavoidable in any serious discussion of redistricting, simple illustrative examples and straightforward percentages or indexes like the disparity ratio not only make the numbers less intrusive but add meaning to the discussion.

Several people were especially helpful in my search for maps and legal information: Ron Grim and Gary Fitzpatrick, at the U.S. Library of Congress; Ted Holynski, at Syracuse University's H. Douglas Barclay Law Library; Congressman James Walsh, of New York's Twenty-fifth Congressional District; Marshall Turner, at the U.S. Bureau of the Census; Daniel Hennessy, of New York State's Legislative Task Force on Demographic Research and Reapportionment; Teresa Neighbor, at the Election Department in Cambridge, Massachusetts; and Douglas Amy, on the political science faculty at Mount Holyoke College. For assistance with imaging software, I am especially grateful to Mike Kirchoff, at the Syracuse University Cartographic Laboratory, and Jeff Bittner, in the Faculty Academic Computing Support Services unit. At home, my wife Marge and daughter Jo were, as always, helpful and tolerant.

1 *Twist and Clout*

GEORGE HERBERT WALKER BUSH, FORTY-first president of the United States, shares a unique political legacy with Elbridge Gerry, our fifth vice president, under James Madison. In 1812, while Gerry was governor of Massachusetts, his party, Thomas Jefferson's Democratic-Republicans, controlled the state legislature. In redrawing senatorial district boundaries after the census of 1810, the Jeffersonians hoped to win more seats by packing Federalist voters into a few strongholds while carving out a long, thin Republican district along the northern, western, and southwestern edges of Essex County. Gerry disliked the plan but signed the remap into law anyway—a veto, he thought, would be improper.[1] The Federalist press was not amused. When a reporter pointed out the new district's

Figure 1.1. The newly configured Essex County, Massachusetts, senatorial district, as embellished by Elkanah Tisdale. Originally published in the Boston *Gazette* for March 26, 1812. From James Parton, *Caricature and Other Comic Art* (New York: Harper and Brothers, 1877), 316.

lizardlike appearance, his editor exclaimed, "Salamander! Call it a Gerrymander!"[2] Artist-cartoonist Elkanah Tisdale added the wings, teeth, and claws (fig. 1.1) that enshrined Gerry's name in the language as a political pejorative, with its hard *g* mispronounced like the *j* in Jerry. Ironically, the sinuous district crafted by the governor's cronies is far less troublesome in form, if not intent, than the cartographic manipulations encouraged by the Department of Justice under the Bush administration. I call them "bushmanders."

This new species is also more ragged around the edge than its nineteenth-century ancestors. In figure 1.2, for instance, New York State's Twelfth Congressional District, crafted as a Hispanic-majority district in 1992, looks more intricate and fragile than the famed Essex County senatorial district of 1812. Although the unadorned Massachusetts prototype provoked cynical slurs, Tisdale's sinister enhancements probably account for its longevity as a political icon: widely reproduced in books and articles on electoral manipulation, the classic gerrymander is rarely rendered as an unembellished contour. By contrast, New York's Twelfth District needs no adornments to explain journalists' delight in labeling it the "Bullwinkle District," after the loquacious moose who shared a Saturday morning spotlight with his

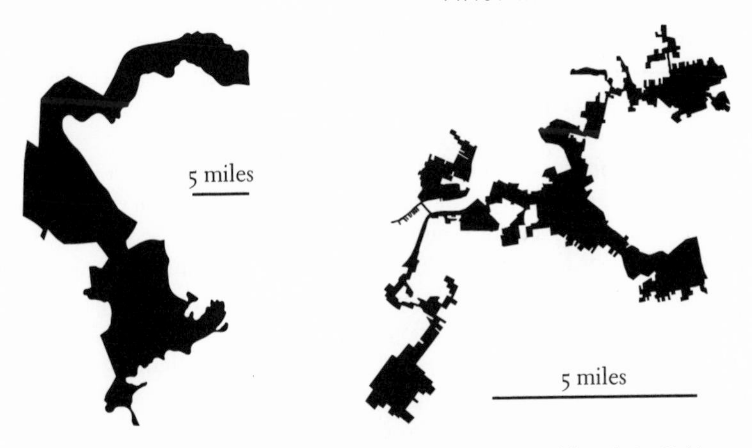

Figure 1.2. Silhouettes of the classic 1812 gerrymander of Essex County (left) and New York's Twelfth Congressional District as configured in 1992 (right). Left-hand map redrawn from map in the Boston *Weekly Messenger,* March 6, 1812, as reproduced in Elmer C. Griffith, *The Rise and Development of the Gerrymander* (Chicago: Scott, Foresman, 1907), 69. Right-hand map compiled from U.S. Bureau of the Census, *Congressional District Atlas: 103rd Congress of the United States* (Washington, D.C., 1993).

cartoon-show sidekick Rocky the flying squirrel.[3] Although the narrow rows of comblike prongs and wider blobs awkwardly connected by thin corridors only faintly resemble antlers, once some clever wag linked the district's contour to the beloved talking moose, the name stuck. Nicknames are rare, though: most bushmanders inspire descriptions like "the 'Z' with drips" (Louisiana's Fourth District), a "spitting amoeba" (Maryland's Third District), and "a pair of earmuffs" (Illinois's Fourth District).[4]

The emergence of these new political critters in the early 1990s is partly a consequence of the Voting Rights Act, passed in 1965 and modified several times. In addition to banning racial discrimination in voter registration, the law defends the right of minority voters to elect candidates of their choice and demands federal scrutiny where past abuse has been especially flagrant. Among other provisions, the Department of Justice must approve the postcensus remap in several states, mostly in the South, as well as the three New York City boroughs containing parts of the Bullwinkle District.[5] Interpreted broadly, the Voting Rights Act also prohibits political cartographers from splitting a district in which members of a minority group constitute a majority.[6] This stricture raises a thorny question: Can the Justice Department deny preclearance when a state chooses not to form a thinly stretched minority-majority district? In rejecting plans submitted by Georgia and North Carolina, George Bush's map editors answered with an assertive "Yes!"[7] Intimidated by earlier rulings as well as eager to accommodate black and Hispanic leaders, New York's mapmakers won preclearance on their first try.[8]

Bushmanders would be difficult, if not impossible, without computers. In Gerry's day, and for more than a century thereafter, the basic building block for congressional districts was the county. Although New England mapmakers eagerly split counties along town lines, political cartographers elsewhere preferred to combine whole counties wherever possible and to split cities, if needed, only along existing precinct boundaries. To explore different configurations, they spread out their maps on a large floor and tallied district populations by hand or by adding machine. Redistricting became more troublesome after the mid-1960s, when the Supreme Court insisted that states not only reconfigure congressional and legislative districts every ten years, as the Constitution intends, but minimize variation among districts in population size. In the early 1990s, with the Civil Rights Division of the Justice Department poised to reject plans that ignored possible minority-majority districts, states turned to interactive computers, electronic maps, and detailed census data, which made it easy to accumulate blocks inhabited by African Americans or Spanish-speaking Americans and link dispersed minority neighborhoods with thin corridors inhabited by few, if any, nonminority "filler people." To ignore this technology was to invite federal judges to draw the lines themselves. After all, Justice officials in Washington had similar tools, as did African American and Hispanic interest groups eager to sue for apparent violations of the Voting Rights Act.

The result typically was a district difficult to describe with maps or words. New York City's Bullwinkle District, for instance, is a polygon with no fewer than 813 sides.[9] In a bill approved by the state legislature and signed into law by the governor in June 1992, District 12's perimeter requires 217 lines of verbal description, which read like the itinerary of a taxi driver trying desperately to run up the meter. Figure 1.3 shows how part of the boundary twisting across Brooklyn helped elect a Hispanic to the House of Representatives by capturing blocks rich in Spanish surnames while avoiding blocks where Hispanics are a minority. The line became law as

> to Linwood street, to Glenmore avenue, to Cleveland street, to Pitkin avenue, to Warwick street, to Glenmore avenue, to Jerome street, to Pitkin avenue, to Warwick street, to Glenmore avenue, to Jerome street, to Pitkin avenue, to Barbey street, to Glenmore av-

> enue, to Schenck avenue, to Liberty avenue, to Barbey street, to Atlantic avenue, to Van Siclen avenue, to Liberty avenue, to Miller avenue, to Glenmore avenue, to Bradford street, to Liberty avenue, to Wyona street, to Glenmore avenue, to Pennsylvania avenue, to Liberty avenue, to Vermont avenue, to Atlantic avenue, to New Jersey avenue, to Jamaica avenue, to Vermont avenue, to Fulton street, to Wyona street....[10]

Lawmakers are word people, and before they vote on a redistricting bill, boundaries composed on a computer screen are converted to verbose lists of street segments, watercourses, and other fixed features.

Because politicians and election officials need to see where voters live, redistricting officials convert the lists back into maps. To supplement the electoral maps of individual states, the Bureau of the Census publishes the *Congressional District Atlas,* a standardized cartographic reference for the fifty states. Because the *Atlas*'s letter-size pages are too small to show on one map the intricate details of computer-crafted gerrymanders, a single district can extend across a dozen pages or more. In the edition for the 103rd Congress, published in 1993, fragments of New York City's Bullwinkle District appear on twenty-two pages.[11] Identified vaguely on the separate, single-page county maps for Brooklyn, Manhattan, and Queens, District 12 crops

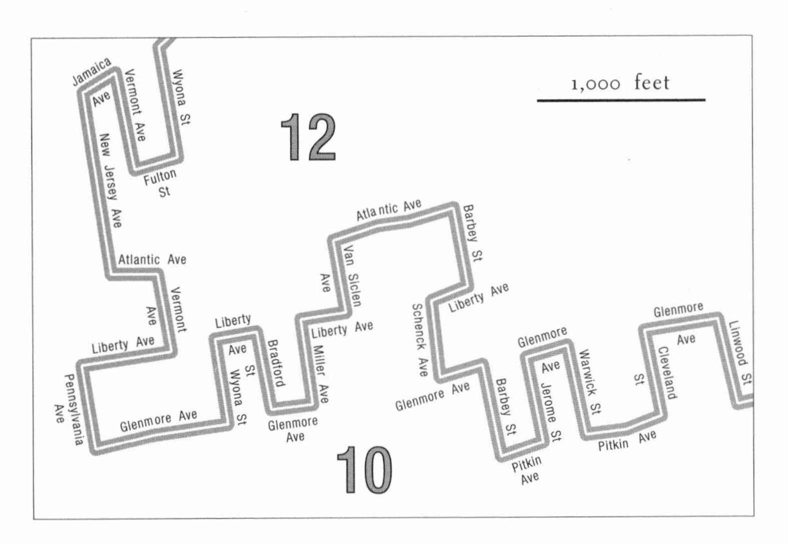

Figure 1.3. Excerpt from U.S. Bureau of the Census, *Congressional District Atlas: 103rd Congress of the United States* (Washington, D.C., 1993), NEW YORK-16.

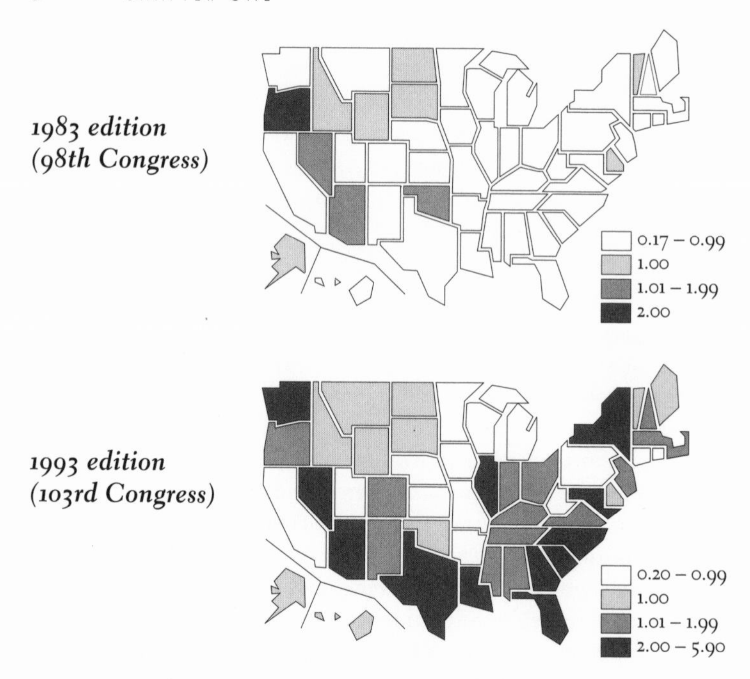

Figure 1.4. Ratio of map pages in the *Congressional District Atlas* to members of the House of Representatives.

up in greater detail on eighteen partial-county inset maps, mostly printed one to a page, at various scales. Figure 1.3, extracted from one of the Brooklyn (Kings County) insets, illustrates the symbols and level of detail. Especially complex portions of the boundary in parts of Queens required eight additional subinset maps—insets of insets—focused on small areas at even larger scales. The cartography of bushmanders is, to coin a word, insetuous.

The *Congressional District Atlas* affords a useful, if idiosyncratic, indicator of the bushmander's radical geometry: the ratio of map pages to House members. Consider New York, which lost three seats after the 1990 census. In 1983 the state's thirty-four congressional districts occupied seventeen pages of maps. Ten years later thirty-one districts required seventy-seven map pages—two and a half pages of maps for every seat. According to the national maps in figure 1.4, for most states the index jumped markedly between the 1983 and 1993 editions. For the earlier year, the highest ratio is an even two for Oregon, where five districts required ten map pages. Only three other states had a ratio over one, and seven states registered equal numbers

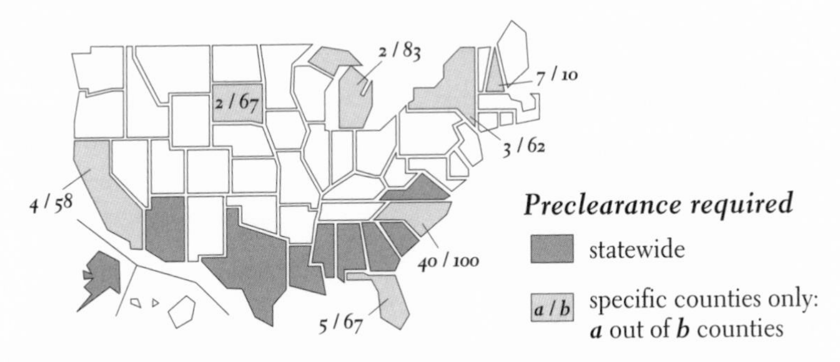

Figure 1.5. States requiring preclearance by the Department of Justice or the U.S. District Court for the District of Columbia. Compiled from *Code of Federal Regulations,* 28 CFR part 51, appendix (7-1-95 edition).

of map pages and House members. By contrast, the map for 1993 shows twenty-five states with a ratio over one, and twelve of these have an index greater than two. Texas, which required 177 map pages to describe thirty districts, had the highest ratio (5.90). Six states were ahead of New York (2.48): Florida (4.52), North Carolina (4.50), Louisiana (4.00), Georgia (3.18), Arizona (2.50), and South Carolina (2.50). Only one of these six, Arizona, had had more map pages than congressional districts ten years earlier.

There's a pattern here: most states with high ratios for 1993 are in the South, most have high relatively large percentages of African American or Spanish-speaking residents, and as figure 1.5 shows, most required preclearance by the Department of Justice. Of these factors, the need for preclearance yields the strongest correlation, however imperfect. Five of the nine states requiring statewide preclearance have ratios of two or higher, as do three of the seven states requiring preclearance for specific counties and only four of the remaining thirty-four states. Among the latter, Illinois (2.30) and Maryland (2.12) designed noncompact districts to elect minority representatives, Washington (2.44) used several complex boundaries in partitioning the Seattle area, and Nevada (2.00), which centered one of its two districts on Las Vegas, threaded the boundary in and out of the city limits.

Close examination accounts for several anomalies. Preclearance is irrelevant to Alaska (1.00) and South Dakota (1.00), each with only one district. California, with fifty-two seats and only four preclearance counties, scored a low ratio (0.94) despite several districts stretching across four or more map pages. More intricate boundaries are also apparent for the South's three full preclearance states with map-seat ratios less than two: between its 1983 and 1993 editions, the *Atlas* registered substantially increased ratios for Alabama (0.43 to 1.71) and Virginia (0.60 to 1.45), which in 1992 elected their first black congressmen since Reconstruction, and for Mississippi (0.40 to 1.40), which redrew its sole minority-majority district to ensure a safer seat for Mike Espy.[12] Although preclearance and racial intent cannot account for all bushmanders, the species flourished in the South, where dispersed populations of rural African Americans thwarted the easy construction of compact minority-majority districts.

Why would a Republican administration favor African American and Latino candidates, almost certain to be Democrats? For the same reason that Elbridge Gerry's Jeffersionian Republicans packed Federalist voters into Essex County's inner district: by creating safe districts in which minority candidates were likely to win, the Bush Republicans added white voters to formerly Democratic districts, which responded, as hoped, by electing Republicans. There was another advantage, though. A widely shared resentment of minority-majority districts, often perceived as yet another affirmative action strike at the prerogatives and values of the white middle-class majority, fueled white dislike for the Democrats' policies and politicians. Because the GOP did not openly advocate minority districts, Republican candidates were free to rail against the bushmanders' flagrantly contorted shapes. Reinforcing the perception of an antiwhite conspiracy were lawsuits filed by aggrieved "filler people" and promptly challenged by pro–civil rights Democrats. Adding to the irony, federal judges appointed by Presidents Reagan and Bush won the public approval of white Republicans by condemning the racial gerrymanders that helped their party take over the House in 1994.

A hypothetical example explains how racial gerrymandering works. The fifty-four squares in figure 1.6 represent counties in a fictitious state accorded three congressional districts in the post-1990 reapportionment. Vying for these seats are two equally fictitious political parties: the Traditionalists and the Modernists. As their name suggests, the Traditionalists salute the flag, love Mom and apple pie, and prefer rural environments, specifically the fifty-one nonshaded cells, representing rural counties, where they outnumber Modernist voters by a slight majority, 81,600 to 71,400. (To simplify the arithmetic, I kept the numbers small, much smaller than the population

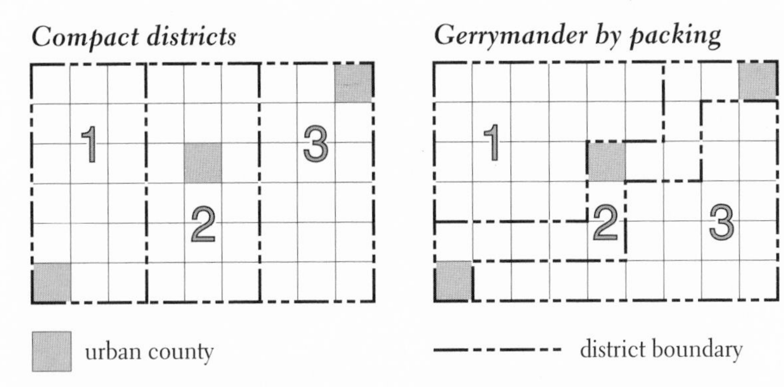

Figure 1.6. Packing opposition votes into a single, comparatively urban district (right) can yield two seats for a largely rural party likely to lose all three otherwise compact districts (left).

now required for three House seats.) By contrast, the Modernists form a substantial majority in the three urban counties, where they outnumber Traditionalist voters 24,000 to 3,000. If all candidates ran at large and every voter had three votes, all three Modernist candidates would win a seat, 95,400 to 84,600. And if the state was partitioned among three equal-sized, equally urban districts, the Modernists would still take every seat.

Numbers and maps explain what's happening. The three comparatively compact districts in the left-hand map yield identical tallies. Within each district, the Modernists reap 1,400 votes from each of seventeen rural counties and 8,000 from the single urban county, for a total of 31,800 votes:

23,800	votes in the seventeen rural counties (17 × 1,400)
+ 8,000	votes in the single urban county
= 31,800	votes for each Modernist candidate

With their slight edge in the countryside more than offset by a weak showing in the city, the Traditionalists can muster only 28,200 votes:

27,200	votes in the seventeen rural counties (17 × 1,600)
+ 1,000	votes in the single urban county
= 28,200	votes for each Traditionalist candidate

With a reverence for compact districts and winner-take-all victories, the Traditionalists might find some solace in an electoral system that reinforces traditional (small *t*) values.

Suppose, though, that their Traditionalist brethren in Washington intervened on behalf of a racial minority of Modernists concentrated in the three urban counties. Suppose that Modernists in the state legislature were encouraged, if not coerced, into a redistricting plan that gave minority-group members the opportunity to elect a "candidate of their choice" by placing all three urban counties in a single district, as in the right-hand map in figure 1.6. Confident that they enjoyed a statewide majority—95,400 to 84,600 would impress most politicians—the Modernists approve the plan and hope for the best. After all, they've courted minority voters over the years, and fair is fair.

But look again at the numbers. In the district linking the three cities, the Modernist candidate—presumably a member of the minority—wins easily. With eleven rural and three urban counties, District 2 is very safe Modernist territory.

15,400	votes in the eleven rural counties (11 × 1,400)
+ 24,000	votes in the three urban counties (3 × 8,000)
= 39,400	votes for the Modernist candidate

The opposing candidate is thoroughly trounced.

17,600	votes in the eleven rural counties (11 × 1,600)
+ 3,000	votes in the three urban counties (3 × 1,000)
= 20,600	votes for the Traditionalist candidate

The other two districts produce different outcomes. Packing Modernist voters into a long, thin district akin in shape and spirit to the Essex County gerrymander yields a pair of highly rural districts in which a Traditionalist candidate enjoys a smaller but nonetheless comfortable majority. With twenty rural counties and no urban counties, each district has 32,000 Traditionalist voters (20 × 1,600) but only 28,000 Modernist voters (20 × 1,400). Able to control the congressional remap, the Traditionalists capture two of the state's three seats and proclaim themselves champions of individual and minority rights.

Meandering boundaries on the left-hand map serve the Traditionalists in two ways: by wasting Modernist votes in District 2 and by diluting Modernist strength in Districts 1 and 3. Racial gerrymanders crafted under the guise of the Voting Rights Act do this too, and so do purely partisan gerrymanders, which are often more subtle in geometry. By adding here and dropping there a clever politician can craft boundaries that confer advantage with little hint of cartographic manipulation. The real test, it seems, is whether the outcome at the polls is substantially better than a party's inherent strength as measured by voter registration or previous elections. In this sense the 1812 manipulation of Massachusetts senatorial districts worked extraordinarily well: although Jeffersonian candidates throughout the state received only 50,164 votes, in contrast to 51,766 votes for their Federalist rivals, Gerry's party won twenty-nine of the Senate's forty seats.[13] Although other districts were drawn to waste Federalist votes, only the outer Essex district was flagrantly misshapen.

Whether a gerrymander is suspicious or subtle often hinges on geography. In major cities like New York and Chicago, where less affluent African Americans inhabit large, generally contiguous but segregated neighborhoods, redistricting officials can usually corral a comfortable majority of black voters without resorting to bizarre boundaries. But even here complex shapes can arise when political cartographers try to maximize the number of black-majority districts or insist on a 60 or 65 percent "supermajority," more certain to elect a minority candidate. By contrast, Hispanic voters, who are less numerous and assimilate more easily, tend to be distributed more widely across the city, in discrete neighborhoods requiring thin corridors like those of New York's Bullwinkle District (fig. 1.2). Carving out even one Latino-majority district can be a delicate task. And in the Deep South, dispersed settlement of African Americans in small towns and rural areas makes black-majority districts difficult if not impossible without long corridors and amorphous appendages. Without computers, a broad interpretation of the Voting Rights Act, and a Justice Department eager to offer minority candidates safe districts, bushmanders would not have appeared.

Despite criticism for encouraging packed minority-majority districts, the Justice Department deserves credit for discouraging cracking, a form of gerrymandering that dilutes a group's electoral clout by breaking up a comparatively compact district with an arguably natural majority. Figure 1.7, in which the hypothetical state's three urban counties are now contiguous, exposes cracking as the antithesis of packing. In the left-hand map, the urban counties form the core of a pro-Modernist district surrounded by a pair of pro-Traditionalist districts. (The numbers are identical to those used to illustrate packing and need not be repeated.) In the right-hand map, boundaries that dismember the Modernist stronghold yield three equally urban districts, all with smaller yet sufficient Modernist majorities. In this case cracking benefits the Modernists, who with these numbers have an ulterior motive for spreading their minority supporters among all three districts. Under other conditions, cracking could help the Traditionalists: if the Modernists held a less solid majority in the urban counties—a more fragile 5,000 to 3,000 edge, for instance—cracking

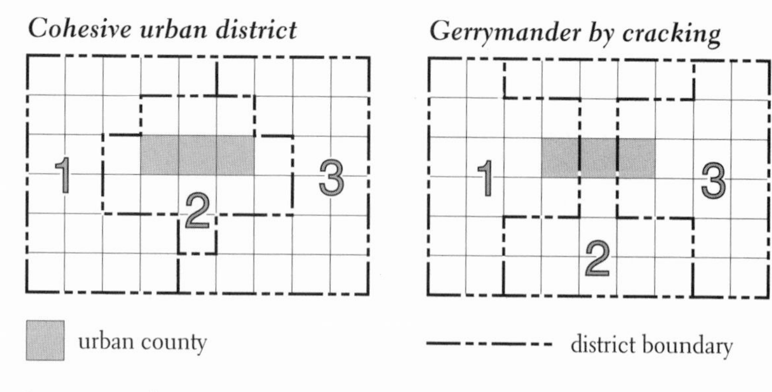

Figure 1.7. Cracking a strongly urban district (left) can dilute the strength of a largely urban party and help its more rural opponents capture all three rural-urban districts (right).

would yield a trio of pro-Traditionalist districts. As noted earlier, maps and numbers are the key to what works and for whom.

Bushmanders are only part of the story. Although racial gerrymanders contributed to a Republican takeover of the House in 1994, the party gained electoral leverage by dismantling gerrymanders devised in the early 1980s, when Democrats enjoyed a firmer hold on state legislatures and the federal judiciary.[14] California Republicans, for instance, were able to undo the redistricting plan that in 1984 helped Democrats win twenty-eight of the state's forty-five House seats with only 48 percent of the statewide vote. Designed by the late Phil Burton, a congressman renowned as a political cartographer, California's 1982 remap wasted Republican votes by cracking as well as packing.[15] In 1994, by contrast, with 52 percent of the statewide congressional vote, Democrats captured only twenty-seven of fifty-two seats. None of the state's new districts merits distinction as a bushmander.[16]

As this book will show, maps have a pivotal role in the decennial realignment of political boundaries. Whether a straightforward description, a tool of manipulation, or an instrument of propaganda, election-district maps link politicians to places, places to voters, and voters to our system of representative democracy. Maps also provide the courts and the media with evidence of compliance or collusion. But as readers will discover, cartographic evidence, especially maps focused narrowly on a district's perimeter, can draw attention away from larger issues of fairness and democratic decision making. Viewed out of context, a voting district's perimeter says little about distance, population distribution, and effective representation.

2 *Gerry's Legacy*

DRAFTED DURING THE SUMMER OF 1787 and ratified over the next three years by all thirteen former British colonies, the Constitution of the United States got right to the point of representative government. Section 1 of Article 1 assigned "legislative Powers" to Congress, "which shall consist of a Senate and House of Representatives," and Section 2 apportioned the House among the states according to their "respective Numbers," to be readjusted every ten years by a census. A formula specifying not more than one representative for each 30,000 persons excluded "Indians not taxed," counted a slave as only three-fifths of a person—the result of these exclusions and reductions is called the apportionment population—and allowed every state at least one House member. Until the first "actual

Enumeration," the number of representatives ranged from one for Rhode Island to ten for Virginia. Outspoken on enumeration and apportionment, the Constitution was silent on congressional boundaries and elections. State legislatures were free to establish districts with one or several members or to have some or all representatives run at large. In 1790 the federal government had little interest in, much less control over, a state's political cartography.

Getting from then to now makes a fascinating story if kept brief and focused. As this chapter observes, the road from minimal to direct federal involvement has few important milestones, mostly after 1960. However meager in number, events like the Civil War, Reconstruction, the Supreme Court's 1964 decision in *Wesberry v. Sanders,* and the Voting Rights Act of 1965 are keys to understanding contemporary congressional geography. Although neither Congress nor the Constitution addresses shape directly, judicial and legislative initiatives to correct discrimination against urban residents and racial minorities paved the way for bushmanders and bizarre boundaries. To illustrate the variety and persistence of Gerry's legacy through the 1980s, the chapter examines the extraordinary reorientations of Mississippi's congressional districts in 1966 and 1984.

If the metaphors "road" and "milestone" are relevant to redistricting, extending the analogy to include a road map seems logical, convenient, and obvious. But it apparently is not obvious to most constitutional historians and political scientists, for my search for a time line summarizing two centuries of congressional apportionment and redistricting turned up only one good example, included here as figure 2.1. I found it in the otherwise dry transcript of a May 1994 voting rights hearing before the Subcommittee on Civil and Constitutional Rights of the House Committee on the Judiciary. Invited to explain the evolution of congressional redistricting, Thomas Durbin, a legislative attorney in the American Law Division of the nonpartisan Congressional Research Service, prepared a chart showing how the nation had arrived at a "crossroads under the Voting Rights Act."[1] I considered redrafting Durbin's sketch but decided to share my serendipitous elation by showing his crude but impressively robust original drawing. Like how-to-get-there maps roughed out daily on napkins and place mats, it works by providing a frame of reference (the decennial censuses) and linking key features (laws and Supreme Court decisions). Durbin went a step further, though, with interpretative comments and labels. In highlighting the shift about 1986 from

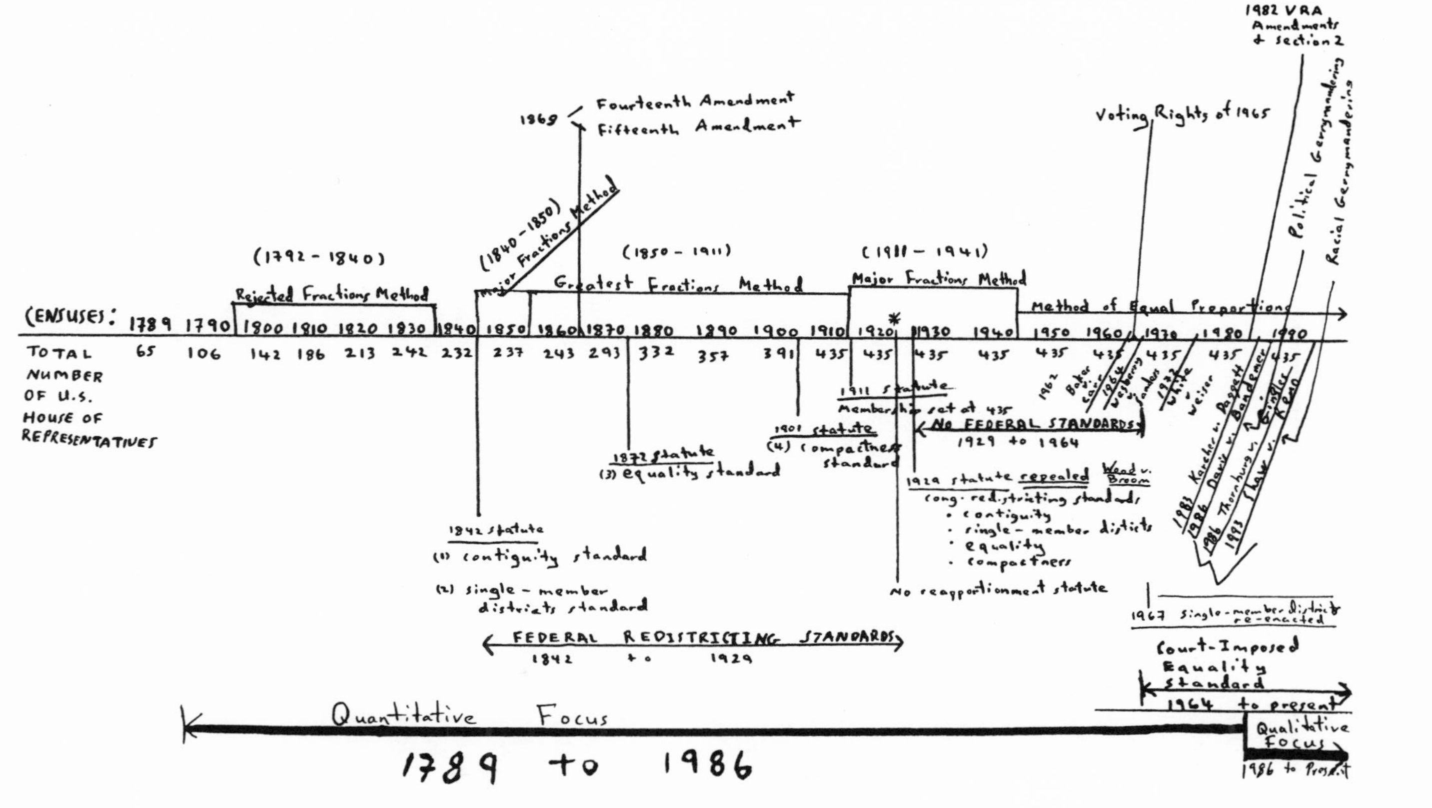

Figure 2.1. History and evolution of congressional reapportionment and redistricting, as sketched by Thomas M. Durbin, May 11, 1994. From U.S. Congress, House Committee on the Judiciary, *Voting Rights: Hearings before the Subcommittee on Civil and Constitutional Rights,* 103rd Cong., 1st and 2nd sess., 1993-94, 87.

a purely quantitative to a more qualitative focus, for instance, he underscored the Supreme Court's increased attention to the representation of minorities.

Durbin's layout not only reveals redistricting as more troublesome than reapportionment but reflects increased hand-wringing since 1960. His horizontal axis, which shows the total number of House members after each enumeration, divides the graph into an upper part addressing apportionment, constitutional amendments, and the Voting Rights Act and a lower portion focusing on redistricting. Too numerous in recent years for the space provided, boundary issues spill over the axis at the far right. By contrast, redistricting was less problematic in earlier years, when states rarely lost seats even though lawmakers raised the ratio of population to representatives after nearly every census since 1790. With one exception, House membership increased in every decade through 1910, when a Congress leery of its cumbersome size capped the number of representatives at 435.

As labels above the axis indicate, Congress changed the recipe for allocating seats several times.[2] From 1792 (when the counts of the first census took effect) through 1840, the "rejected fractions" method divided each state's apportionment population by a fixed ratio to obtain an exact "quota," consisting of a whole number and a fraction. After giving a representative to each state with a quota less than one, the method earned its name by considering only the whole numbers. For example, division by 40,000, the ratio Congress set in 1822 for the post-1820 reallocation, yields only one representative for an apportionment population of 79,000 but two seats for a state with 81,000. Not surprisingly, states just shy of the threshold considered the method unfair. In 1842 Congress addressed these objections by adopting the "major fractions" method, which awarded an additional representative for any fraction over one-half.[3] Although a state with an apportionment population of 61,000 might have welcomed the change, a neighbor with only 59,000 residents would hardly have been enthusiastic.

Table 2.1 "Greatest fractions" method of apportionment

State	Population	Exact quota	Whole number	Seats
A	453,320	45.332	45	45
B	443,310	44.331	44	44
C	103,370	10.337	10	11
Total	1,000,000	100.000	99	100

In 1852, at the urging of Ohio congressman Samuel Vinton, lawmakers adopted the "greatest fractions" method, which fixed the number of representatives and hence the ratio used to compute each state's quota. After assigning representatives according to the whole-number portion of the quota—with at least one seat to every state—Vinton's recipe ranks the fractions of all states below their quota and uses these ranks to allocate remaining seats. If this is confusing, consider table 2.1, which describes the method for a hypothetical nation with three states, a total population of 1 million, 100 representatives, and an apportionment ratio of 10,000.[4] In this example, the whole numbers account for only 99 seats, and state C, with the greatest fraction (0.337), receives the one hundredth representative. Fair enough, I suppose: the extra seat must go somewhere.

If Vinton's method seems foolproof, think again. Setting total membership at 101, which lowers the apportionment ratio to 9,901, actually reduces the number of seats assigned state C. What's worse, this reduction could occur even if C's population rose slightly relative to the other states' populations. If this seems implausible, if not preposterous, look at table 2.2.

With the total number of representatives increased by one, all states register higher quotas. But because C has the smallest fraction, A and B divide the two remaining seats. Apportionment experts call this the Alabama paradox because the 1890 census would have given Alabama eight seats in a House of 299 members but only seven seats if membership was fixed at 300.[5] C. W. Seaton, chief clerk of the U.S. Census Office, discovered this idiosyncrasy when computing apportionments for all House sizes between 275 and 350 seats. Useful to lawmakers eager to avoid controversy, Seaton's trial calculations informed the decision to increase House membership to 325.

Seaton's discovery was not the only aberration. In 1901 the House Select Committee on the twelfth census noticed that Maine would gain a seat in a House fixed at 386 members while Virginia, which had

Table 2.2. Illustration of the "Alabama paradox": Increasing the total number of seats by one reduces the number of seats for state C.

State	Population	Exact quota	Whole number	Seats
A	452,170	45.669	45	46
B	442,260	44.668	44	45
C	105,570	10.663	10	10
Total	1,000,000	101.000	99	101

grown more rapidly, would lose a seat.[6] But because neither state was growing as rapidly as the country as a whole, Maine's quota declined by 1.3 percent (from 3.595 to 3.548) while Virginia's dropped by 0.9 percent (from 9.599 to 9.509). Note the change in fractions: despite a proportionally lower quota, Maine (now with a fraction of 0.548) had gained an advantage over Virginia (with only 0.509) for an extra seat. Called the population paradox, this phenomenon reveals a flaw in awarding extra seats according to remainders, or fractions, that ignore the relative sizes of states. A third anomaly, dubbed the new states paradox, occurs when the admission of a new state with its fair share of representatives affects allocations to other states.

In 1911, at the urging of Cornell University philosophy professor Walter Wilcox, Congress returned to major fractions for the post-1910 reapportionment.[7] With the House now fixed at 435 seats, the 1911 law raised the apportionment ratio slightly to avoid the four extra seats that would have resulted from merely dividing the apportionment population by 435. Although major fractions also guided the post-1930 reapportionment, the 1920 census threw Congress into endless debate by revealing extraordinary growth in large cities and a marked decline in rural areas. Forty-two reapportionment bills were introduced in seven years, but none passed, as passive-aggressive rural politicians prevailed and the House skipped its mandated post-1920 realignment.[8]

Congress altered the recipe once again in 1941 by adopting the method of "equal proportions," which uses a one person, one vote standard to round fractions upward or downward. Developed in 1911 by Census Bureau statistician Joseph Hill and promoted vigorously by Harvard University mathematics professor Edward Huntington, equal proportions addresses relative size by favoring states with smaller populations and fewer members.[9] Population size is important because residents of a state with a quota of 1.498 and only one member are less equitably represented in the House than residents of a state with a quota of 60.499 and sixty members.

To promote equality in the average number of constituents per representative, Hill's recipe gives more weight to the fractions of smaller states. It does this by calculating a "priority value" for each state for each seat being claimed. A state's priority value is computed by dividing its apportionment population by $1/[N(N-1)]^{1/2}$, where N is 2 for a state trying to claim its second seat, 3 for a third seat, and so on. Since each state automatically receives one representative, the

Table 2.3. Post-1990 allocation of the first and last six seats according to the method of equal proportions

Numbered seat in the House	State	Numbered seat in the state	Priority value
First six seats			
51	California	2	21,099,536
52	New York	2	12,759,392
53	California	3	12,181,823
54	Texas	2	12,063,104
55	Florida	2	9,194,765
56	California	4	8,613,850
Last six seats			
430	Mississippi	5	578,346
431	Wisconsin	9	578,265
432	Florida	23	678,070
433	Tennessee	9	577,075
434	Oklahoma	6	576,497
435	Washington	9	576,049
Three seats beyond the 435th			
436	Massachusetts	11	574,847
437	New Jersey	14	574,367
438	New York	32	572,914

Source: Condensed from Dudley L. Poston Jr., "The U.S. Census and Congressional Apportionment," *Society* 34 (March–April 1997): 39.

recipe starts with the fifty-first seat and awards all subsequent seats separately, to the state with the highest priority value for that seat.

Confused? Look carefully at table 2.3, which describes the process with a few examples from the post-1990 reapportionment. The first six rows report the winning priority values for House seats 51 through 56. California, the largest state, captured the fifty-first seat because its second-seat priority value (21,099,536) was larger than New York's (12,759,392). Once assigned a second seat, California based its claim for the fifty-second seat on its third-seat priority (12,181,823), which was lower than New York's. Although New York won the fifty-second seat, California captured the fifty-third seat because its third-seat priority exceeded the second-seat priorities of Texas (12,063,104) and Florida (9,194,765), sufficiently large nonetheless to take seats 54 and 55. Even so, the Golden State's substantial fourth-seat priority (8,613,850) beat other states' claims to seat 56.

The next six rows report the winning priority values for the last six

House seats, and the final three rows show the next three highest priority values, with which Massachusetts, New Jersey, and New York narrowly lost an eleventh, a fourteenth, and a thirty-second seat, respectively. Near misses like these explain why large urban states with a substantial minority population eagerly lobby Congress and the Census Bureau to reduce or adjust for the undercount.

For the nation's first fifty years, federal lawmakers ignored redistricting entirely: as long as a state elected no more than its allotted number of representatives, its legislature could draw district boundaries any way it wanted. When Washington finally acted, in 1842, it was only to write into the post-1840 reapportionment bill a requirement for contiguous, single-member districts. Although the new law reflected the widely held opinion that multimember districts and at-large elections can concentrate too much power in too small a part of the state, some leaders considered the statute an unwarranted, if not unconstitutional, encroachment on states' rights. In signing the measure, President John Tyler acknowledged "deep and strong doubts" about the power of Congress "to command the States to make new regulations or alter their existing regulations."[10] John Quincy Adams, the former president who returned to Congress for seventeen years, was not so sanguine. Denouncing the law as "pernicious in its immediate operations, and imminently dangerous in its tendencies," Adams introduced a House resolution protesting Tyler's defense of the bill.[11] Despite repeated attempts at passage, Adams's resolution never achieved the necessary two-thirds vote.

Durbin (fig. 2.1) shows the contiguity and single-member mandates as launching an era of "federal redistricting standards," but the label and bold line on his diagram overstate their impact. The post-1850 reapportionment bill said nothing about districts, for instance, although corresponding statutes for the censuses 1860 through 1910 endorsed the contiguity and single-member standards. In 1872 Congress added a third standard, prescribing districts with "as near as practicable, an equal number of inhabitants."[12] Retained for post-1880 and post-1890 reapportionment laws, the equality standard was joined in 1901 by the equally vague compactness standard, which called for a "contiguous and compact territory and containing as nearly as practicable an equal number of inhabitants."[13]

The four-point redistricting standard was short-lived if not inconsequential. Although the 1911 apportionment bill reaffirmed all four requirements, Congress failed to reapportion itself for the 1920 census, and the 1929 reapportionment law, passed before the 1930 census to make reallocation certain if not painless, said nothing about contiguity, single-member districts, equality, or compactness.[14] In 1932, moreover, in *Wood v. Broom,* the Supreme Court retroactively wiped out federal redistricting standards by ruling that the provisions of a reapportionment statute did not apply automatically and indefinitely to subsequent remaps.[15] Because the new law effectively repealed the previous (1911) statute, federal redistricting standards ended in 1929.

Wood v. Broom started out as *Broom v. Wood,* in 1932, when would-be congressional candidate Stewart Broom sued Mississippi's secretary of state Walker Wood and other officials over the recent remap. In dividing the state among seven districts, one fewer than its post-1910 allotment, the Mississippi legislature had assigned 414,000 residents to one district but only 184,000 to another. Broom charged that the new map violated the 1911 reapportionment statute as well as Article 1, Section 1 of the Constitution, and he asked the Federal District Court for Southern Mississippi to prohibit use of the new boundaries.[16] In finding for the plaintiff, two of the three judges hearing the case agreed that "it would have been possible to divide the state into seven districts...having approximately the same number of inhabitants."[17] True but irrelevant, asserted the third judge, who considered his colleagues' injunction "an encroachment by the court upon the legislative power of a state."[18] Encouraged, state officials appealed to the Supreme Court, which overturned the district court's injunction because the equality standard, as well as other provisions of the 1911 statute, no longer held.[19]

Mississippi's unequal districts were hardly unique. As the dissenting jurist observed, the 1930 populations of districts drawn up by other state legislatures varied from 250,000 to 450,000 in Alabama, from 165,000 to 350,000 in California, from 225,000 to 400,000 in Michigan, from 168,000 to 634,000 in Ohio, from 125,000 to 445,000 in Pennsylvania, from 200,000 to 650,000 in South Dakota, and from 195,000 to 380,000 in Tennessee.[20] What's more, substantial inequalities were apparent in the post-1900 remap, in which the Mississippi legislature carved out districts with populations ranging from 162,000 to 232,000.[21] With the courts reluctant to intervene, whatever federal redistricting standards were thought to exist made little impression on state lawmakers.

Scandalous at the beginning of century, disparities between the states' smallest and largest districts grew more pronounced during the 1920s and 1930s, as rural migrants streamed into New York, Chicago,

Ratio of largest to smallest district populations, Census of 1900, 58th Congress (1903)

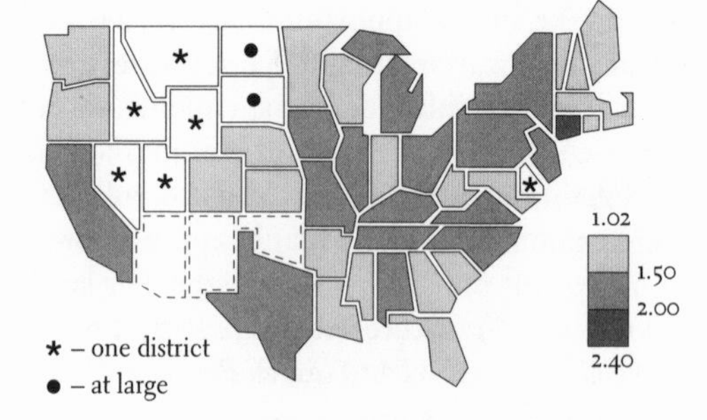

Ratio of largest to smallest district populations, Census of 1950, 88th Congress (1953)

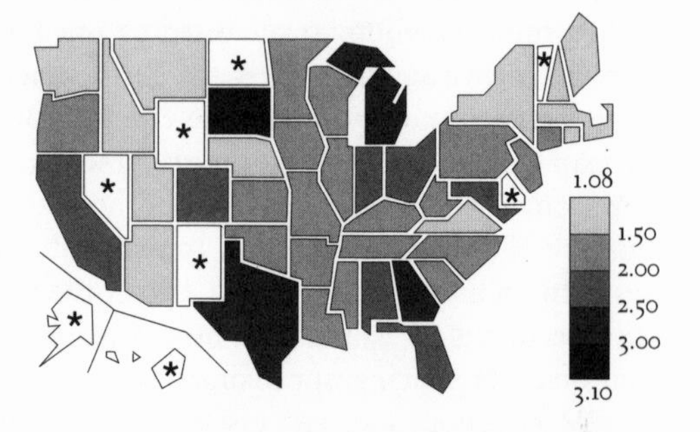

Figure 2.2. Ratio of the populations of states' biggest and smallest congressional districts, for the reapportionments following the censuses of 1900 and 1950. Compiled from data in Stanley B. Parsons, Michael J. Dubin, and Karen Toombs Parsons, *United States Congressional Districts, 1883-1913* (New York: Greenwood Press, 1990), 309-439; and U.S. Bureau of the Census, *Congressional District Data Book (Districts of the 88th Congress)—A Statistical Abstract Supplement* (Washington, D.C., 1963), 560.

and other large cities.[22] Mississippi reflected the national pattern: in 1900 the Magnolia State's largest congressional district had 1.4 times as many residents as its smallest district, and by 1950 the disparity ratio had risen to 1.6. As figure 2.2 shows, between 1900 and 1950 the national norm had passed 1.5, and the number of states with a disparity ratio greater than 2 had risen from one to eleven. Most egregious was South Dakota, with district populations ranging from 159,000 to 493,000. Legislatures either failed to redistrict—political scientists call this "silent gerrymandering"—or drew new boundaries that favored rural constituencies.[23]

Legislative districts were more lopsided than their congressional counterparts. As of 1962, the disparity ratio for the lower houses of state legislatures ranged from 2.2 for Hawaii to 1,081.3 for New

Hampshire, one of four states that had not redistricted since 1900.[24] Especially egregious disparities also occurred in Vermont, where the largest lower-house district had 987 times as many residents as the smallest district, and Connecticut and Florida, with ratios of 424.5 and 108.7, respectively. Moreover, a recent remap was no guarantee of equality: of the twenty-seven states with disparity ratios greater than 10, nine had redistricted between 1951 and 1960 and another eight had redrawn their boundaries the previous year. Districts dominated by rural and small-town residents were generally smaller and more numerous than districts representing large-city and suburban residents, who had less clout in allocating state funds as well as in redrawing congressional boundaries. Justifications for perpetuating this imbalance included the assumed superiority of people living close to the land and the need to protect agricultural interests from the numerical tyranny of industrialists, labor unions, and other greedy city folk.[25]

Although Durbin extends the era of "no federal standards" through 1964, the Supreme Court signaled an end to rural dominance two years earlier, in its landmark decision in *Baker v. Carr.* Not yet willing to meddle with congressional districts, the high court asserted its authority, under the Fourteenth Amendment's "equal protection" clause, to judge the fairness of a recent legislative remap.[26] A reading of the amendment suggests the ruling was long overdue: titled "citizenship rights not to be abridged" by its Reconstruction-era authors, the measure guarantees that "no State shall...deny to any person within its jurisdiction the equal protection of the laws."[27] At issue were Tennessee's general assembly districts, last reconfigured in 1901. In approving a challenge by the mayor of Nashville, the justices triggered malapportionment lawsuits across the country.[28]

By 1964 the Supreme Court was willing to look at congressional districts. In *Wesberry v. Sanders,* a group of Georgia citizens who challenged the constitutionality of districts last redrawn in 1931 cited 1960 census figures showing that the largest district held three times as many people as the smallest. In venting his colleagues' outrage, Justice Hugo Black opined that "while it may not be possible to draw congressional districts with mathematical precision, that is no excuse for ignoring our Constitution's plain objective of making equal representation of equal numbers of people the fundamental goal for the House of Representatives. That is the high standard of justice and common sense which the Founders set for us."[29]

The Court grew increasingly intolerant of numerical inequality.[30] In 1969, in *Wells v. Rockefeller* (not shown in Durbin's diagram), the justices rejected a New York congressional redistricting plan with a disparity ratio of 1.139.[31] And in 1973, in *White v. Weiser,* the Court rejected a Texas plan with a disparity of 1.042.[32] Even so, a good explanation might make a minor imbalance acceptable, at least for legislative districts. In 1973, for instance, in *Mahan v. Howell,* the justices accepted a disparity ratio of 1.18 for Virginia's house of delegates because "with one exception, the delegate districts followed political jurisdictional lines of the counties and cities."[33] By contrast, the federal courts have generally applied stricter standards to congressional remaps.

Numerical equality still trumped race. In 1983, for instance, in *Karcher v. Daggett,* the Supreme Court rejected New Jersey congressional districts with a disparity ratio of only 1.007. The legislature, it seems, might easily have lowered the ratio to 1.005 at the expense of African American constituencies. In writing for the 5-4 majority, Justice William Brennan agreed with a lower court that the state's plan "was not a good-faith effort to achieve population equality."[34] Reluctant to reverse a lower-court ruling, the high court found the legislature's interest in preserving minority voting strength less significant and compelling than the constitutional mandate of numerical equality.

Federal judges were not the only activists. Caught up in the spirit of the civil rights movement, Congress passed the Voting Rights Act of 1965. In a belated effort to enforce the ninety-five-year-old Fifteenth Amendment, which had yet to fulfill its title "race no bar to voting rights," the new law outlawed poll taxes, literacy tests, and similar obstacles to minority voting.[35] Two provisions addressed gerrymandering, at least indirectly. Section 2, known as the nondilution provision, prohibits discriminatory voting laws, including election districts that dilute minority voting strength. By contrast, Section 5, the nonretrogression provision, requires that the Department of Justice or the Federal District Court for the District of Columbia approve any changes in voting laws or procedures in areas with a history of racial discrimination.[36] The results were impressive, especially in Mississippi, where registration among voting-age blacks increased from less than 7 percent in 1964 to more than 62 percent in 1971 and leading candidates no longer campaigned as segregationists.[37]

The 1965 law was just the beginning. In 1967 Congress banned at-large congressional districts, which had tended to weaken minority candidates and diminish the clout of minority voters.[38] In 1970 federal lawmakers renewed the Voting Rights Act for five years, and in 1975 and 1982 they extended its life seven and twenty-five years, respectively.[39] Congress also expanded the law's scope by granting protection to "language minorities" (Asian Americans, Hispanics, and Native Americans) in 1975 and removing the need to prove discriminatory intent in 1982.[40] In shifting the focus from ballot boxes to boundaries, the 1982 amendment heralded an increase in minority officeholders.

Four years later, in *Thornburg v. Gingles,* the Supreme Court confirmed that plaintiffs need not prove intent to discriminate. Whatever a state or local legislature's goal, election district boundaries that dilute the voting strength of minorities violate the Voting Rights Act as well as the Fourteenth and Fifteenth Amendments.[41] State officials and the Justice Department saw *Gingles* as a mandate for delineating minority-majority districts wherever, in the Court's words, a "politically cohesive" minority group was "sufficiently large and geographically compact to constitute a majority in a single-member district" but unable to elect its preferred candidates because of racial bloc voting.[42] The threat was clear: if a remap didn't measure up, the federal judiciary could appoint a consultant, or "special master," to draft its own boundaries. In another 1986 landmark case, *Davis v. Bandemer,* the justices agreed to hear challenges to partisan gerrymandering of congressional districts.[43] By refocusing political cartography on "qualitative" issues (fig. 2.1) more complex than numerical equality, the high court laid a foundation for the bizarre shapes and contentious litigation of post-1990 redistricting.

Back to Mississippi, the undisputed poster child of the Voting Rights Act. Few states have practiced such cartographically blatant segregation, and few have been so effectively reformed, geographically as well as politically, by Congress and the federal courts. To describe what happened and why, I juxtaposed six maps (fig. 2.3) showing successive configurations of the state's congressional districts. My maps span the mid-1950s through the very early 1990s, when new census results triggered a further remap. Thinner boundary lines describe counties, while heavier boundaries delineate congressional districts, identified by number. A light gray shading highlights counties with an African American majority in 1960 (top row) or 1980 (bottom row). Although Mississippi has proportionately more black residents than any other state, the percentage (42 percent in 1960, 35 percent in 1980) has fallen steadily since 1900, largely through migration to Chicago and other northern cities. Despite continued losses, black majorities remain prominent in the flat, fertile

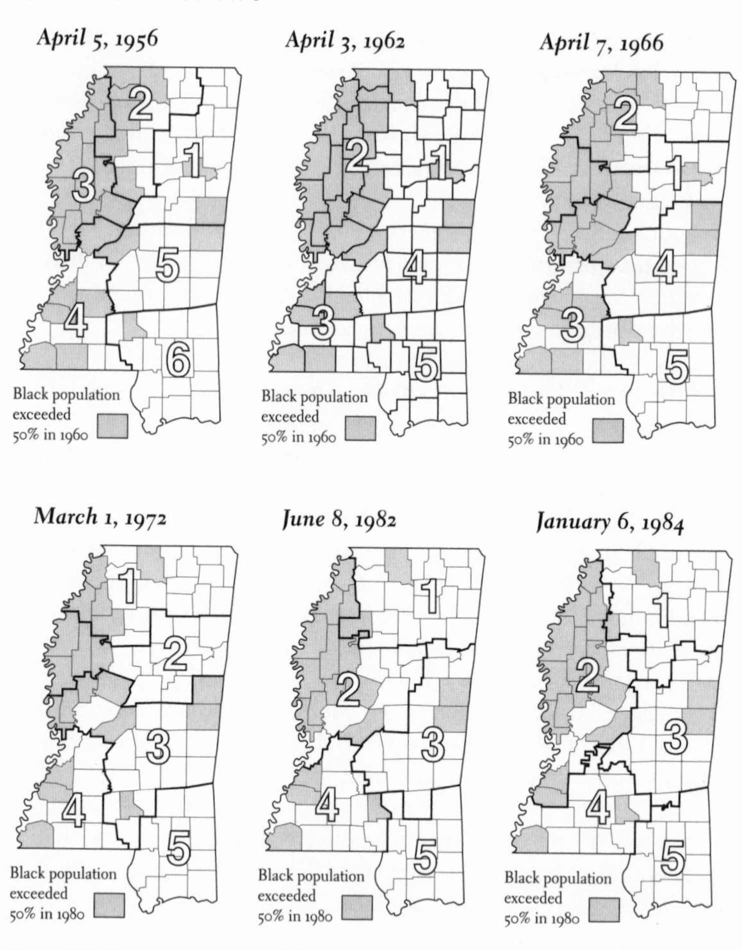

Figure 2.3. Mississippi congressional districts, by date of establishment, for the Eighty-fifth through the 102nd Congresses. Compiled from U.S. Bureau of the Census, *Congressional District Atlas*, various years.

northwestern part of the state known as the Delta, famous for its cotton plantations, antebellum mansions, and economically enslaved tenant farmers.[44]

Until 1962 the Delta was a distinct congressional district, with a largely disfranchised black majority and borders essentially unchanged since 1882.[45] As described in the upper left map of figure 2.3, District 3 had survived a major redistricting in 1952, when persistent out-migration and the loss of a House seat forced revision of boundaries established in 1932, as well as a minor adjustment in 1956, when the legislature merely transferred a single county between two other districts.[46] After out-migration cost the state another seat in 1962, the legislature combined the Delta with District 2, directly east, while letting the other districts stand intact (fig. 2.3, upper center). Expedient consolidation made sense to segregationist governor Ross Barnett and like-minded legislators: the two districts led the state in lost population, and their merger would force the Delta's moderate congressman into a difficult (and losing) race with a more traditional colleague.[47] Although consolidation increased the disparity ratio from 1.93 to 2.06, Mississippi's malapportionment was hardly exceptional.[48] More important to white conservatives, although African Americans made up only 59 percent of the new district's total population—down from 66 percent for old District 3—only 52 percent were of voting age and far fewer were registered.

Four years later new threats to segregation inspired a radically different remap. Not only was the disparity ratio too high to survive a constitutional challenge in the wake of *Wesberry v. Sanders,* but the Delta's blacks, empowered by the Voting Rights Act, might well elect another moderate to Congress or, worse yet, one of their own.[49] Eager to avoid boundaries drawn by a court-appointed cartographer—no telling what a special master might propose—the legislature divided the Delta among three east-west trending districts (fig. 2.3, upper right), none with a black majority. The new plan addressed the one person, one vote imperative by reducing the disparity ratio to 1.06.[50] In 1972, when population shifts and a judicial mandate for fuller numerical equality dictated another remap, the legislature imposed a similar east-west alignment (fig. 2.3, lower left) with an even lower disparity ratio, 1.04.[51]

Undaunted by federal laws and landmark Supreme Court decisions, segregationist politicians shifted tactics. In *Black Votes Count: Political Empowerment in Mississippi after 1965,* Frank Parker, who worked as a civil rights lawyer in Jackson from 1968 to 1981, lists twelve acts of "massive resistance" by the state legislature, which in 1966 alone not only manipulated congressional and legislative districts but took the post of local school superintendent off the ballot and adopted at-large voting for county officials and school board members.[52] When civil rights organizations objected in court, the state's lawyers mounted a spirited defense, which Mississippi-bred federal district judges almost always agreed with.

Especially discouraging was the Supreme Court's 1967 decision, in

Connor v. Johnson, upholding the 1966 congressional boundaries, which the Mississippi Free Democratic Party had challenged as a violation of the Fourteenth and Fifteenth Amendments.[53] In a dismissive, one-sentence summary judgment—Parker called the justices' decision not to hear oral arguments a "nondecision"—the high court tacitly agreed with the federal district judges who had accepted the state's 1.06 disparity ratio and rejected the MFDP's proof of racial intent.[54] The Court's silence had a further effect in 1971, when Mississippi forwarded its post-1970 congressional remap to Washington for Section 5 preclearance and the Department of Justice deferred to the *Connor* decision.[55] According to Parker, approval of the Delta gerrymander also weakened the morale of attorneys planning other challenges to legislative intransigence.[56]

Can shape reveal anything about intent? Not in the minds of the Hinds County officials who in 1973 pushed racial gerrymandering to the limit with the five county-supervisor districts in figure 2.4. Their goal, the county's attorneys argued in *Kirksey v. Board of Supervisors of Hinds County,* was equality in both population and county highways, not (as several black voters charged) the partition of Jackson's African American community among all five districts.[57] Even though the district judge agreed with the plaintiffs that District 3 resembled a turkey and District 4 a "baby elephant," he sided with the county, and a three-judge panel of the United States Court of Appeals, Fifth Circuit, agreed.[58] Convinced that the map spoke for itself, the plaintiffs appealed to the entire Fifth Circuit, which ruled 10-3 that the boundaries unconstitutionally "fragment[ed] a geographically concentrated but substantially black minority in a community where bloc voting has been a political way of life."[59] In sending the case back to the district court for resolution, the appeals court observed that "there was simply too much emphasis on the administrative convenience of equal road and bridge responsibility at the expense of effective black minority participation in democracy."[60]

Blatant gerrymandering at both state and local levels highlighted a critical weakness of the Voting Rights Act: the need to demonstrate racial intent. When the law came up for renewal in 1982, Congress shifted the issue of proof from intent to results: to demonstrate a violation of Section 2, a plaintiff need only show that "based on a totality of circumstances...members [of a minority group] have less opportunity than other members of the electorate to participate in the political process and to elect the representatives of their choice."[61] In addition to simplifying court challenges, the amended law gave added

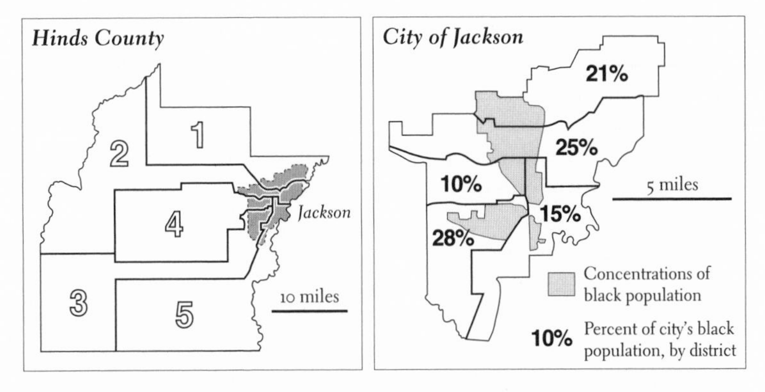

Figure 2.4. Boundaries adopted in 1973 by the Hinds County (Mississippi) Board of Supervisors divided Jackson's black population (right) among all five districts (right). Compiled from Frank R. Parker, *Black Votes Count: Political Empowerment in Mississippi after 1965* (Chapel Hill: University of North Carolina Press, 1990), 155-56.

clout to the Department of Justice, which now had stronger grounds for denying Section 5 preclearance as well as less justification for neglecting its responsibility.[62] A state or locality now had to prove that any changes were not discriminatory, whereas the old standard accepted revised districts or procedures no worse than those they replaced. Fortunately for Mississippi civil rights activists, federal watchdogs were now up to the task. Although deliberately ineffective enforcement during the Nixon administration had undermined earlier, weaker versions of the law, the Department of Justice reorganized its Voting Section in 1976 and established a Section 5 unit to review changes in district boundaries and voting procedures.[63]

It was Washington's turn to resist. In 1981 Mississippi submitted a post-1980 plan perpetuating a divided Delta.[64] Although state officials claimed the remap was less discriminatory than their 1972 plan, the Justice Department denied preclearance because the proposed boundaries diluted the strength of the Delta's black voters. Fearing an impasse might continue the 1972 districts through the 1982 elections, black voters in Greenwood presented the district court with two alternative plans that kept the Delta intact in a district where either 64 or 66 percent of residents were African American. A legislator offered a third plan, backed by the AFL-CIO, in which blacks composed not quite 54 percent of the population of a differently configured Delta district. Conceding that the 1972 plan was now unconstitutionally imbalanced—1980 census tabulations indicated a disparity ratio of

1.19—the court adopted the latter plan (fig. 2.3, lower center) as an interim solution consistent with past practices. "Establishment of a 'safe' minority seat is not a federal prerequisite of a reapportionment plan," the judges noted, "and thus the state's policy of favoring two 40 percent or better black population districts rather than one district with a large black population was legitimate."[65] That the seat was not "safe" was clear when the Democratic candidate, a black, lost by three thousand votes to a conservative white Republican.[66]

Black voters fared better the following year, when the Supreme Court rejected the stopgap map and sent the case back to the district court. Although consistent with state policy, the new districts did not satisfy the Voting Rights Act Amendments of 1982, signed by President Reagan less than four weeks after the district court imposed its temporary solution. Because the law now focused on winning elections, a Delta district in which African Americans constituted 48.05 percent of the voting-age population warranted more careful cartographic experimentation. The district judges complied by crafting a new map, with an obsessively small disparity ratio of 1.0004 and a Delta district in which 52.8 percent of voting-age residents were black.[67] Although the white Republican incumbent again beat his black Democratic challenger in 1984, a new candidate, black attorney Mike Espy, captured the seat in 1986 by a slim majority.[68] Mississippi's first African American House member since Reconstruction, Espy demonstrated the power of incumbency by decisive victories in later contests.

Mississippi's court-imposed congressional districts hint of radical changes in political cartography. Close inspection of the 1982 map (fig. 2.3, lower center) reveals boundaries cutting across two counties: a necessary compromise for the mapmaker pursuing equal populations as well as a black majority in the Delta. And with even more intricate borders subdividing eight counties, the 1984 map (fig. 2.3, lower right) achieved a black voting-age majority in the Delta and district populations that differ by no more than 218 persons. Although the Delta's naturally compact black majority had no need for the abstruse appendages of a bushmander, locally intricate borders were often the only way to blend minority empowerment with strict numerical equality.

3 *Thin Majorities*

IF PRESSED FOR A SIMPLISTIC EXPLANATION of bushmanders, I'd be tempted to blame air conditioning and computers. Our ability to control indoor climate accounts for much of the recent migration of jobs and retirees to the Sun Belt—movements that triggered substantial reallocations of House seats and laid a foundation for Republican victories in statehouses throughout the South. And computers promoted the obsessive tinkering with boundaries and numbers needed to maximize minority representation as well as ensure strict numerical equality. These days, though, no self-respecting social scientist would stoop to such flagrant technological determinism, especially after fundamental changes in the political climate clearly eclipsed the contributions of refrigeration and electronics. As

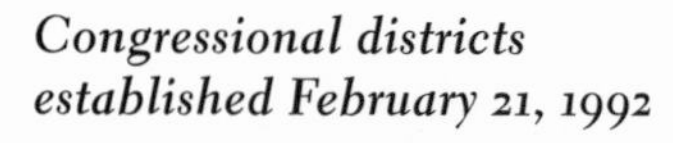

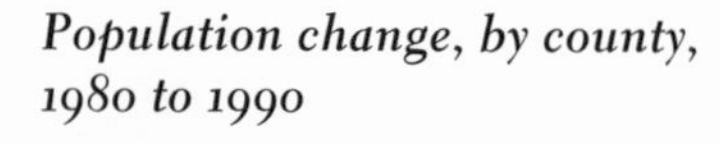

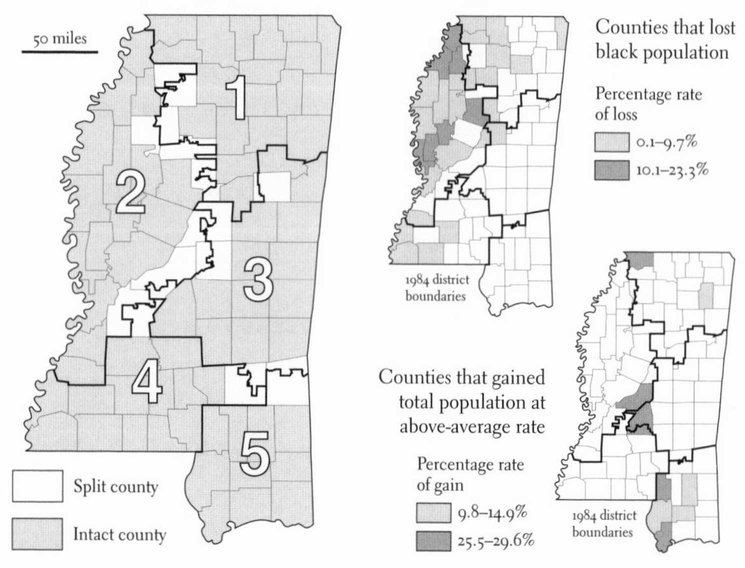

Figure 3.1. Mississippi's post-1990 congressional remap (left) compensated for black losses in the Delta (upper right) and slow growth or loss of total population throughout much of the state (lower right). Compiled from U.S. Bureau of the Census, *Congressional District Atlas: 103rd Congress of the United States* (Washington, D.C., 1993).

this chapter observes, redistricting in the 1990s was also far more than a tale of southern mores, civil rights, and resistance overcome. It's a story that reaches into the metropolitan North, mixes race with old-fashioned partisan politics, and exposes a Supreme Court deeply divided on the meaning of compactness and the need for remedial racial gerrymandering.

The story is conveniently picked up in Mississippi in the early 1990s, when state Democrats controlled the legislature, the governorship, and post-1990 redistricting.[1] Confident of surviving the November 1991 elections, legislative leaders turned the task of redistricting over to a committee, which was unable to agree on a remap during the 1991 session but completed its work shortly thereafter and scheduled a vote early in 1992.[2] Debate would be brief, party officials hoped, and Democratic governor Ray Mabus would quickly approve the bill and forward the new boundaries to Washington for preclearance. Or so they planned. In an upset, Mabus lost to Republican Kirk

Fordice.[3] When inaugurated in January, Fordice could easily thwart the dual goal of satisfying the Voting Rights Act while ensuring the reelection of all five Democratic incumbents. Aware of the threat, Mabus convened a special session of the legislature on December 18. After three days of discussion, the legislature approved the bill, which Mabus signed the same day. Two months later the Justice Department made the new boundaries official (fig. 3.1, left).[4]

Preclearance was never in doubt: having learned the hard way eight years earlier, Mississippi politicians were unwilling to risk letting judges choose their boundaries. And because state officials had no interest in radically altering the basic shapes and orientations of the 1984 map, the legislature's redistricting committee might have finished months earlier had black leaders not objected to early drafts. The problem was that Delta residents, largely black, were not only voting but voting with their feet: between 1980 and 1990 the population of District 2 dropped by 6.2 percent, in contrast to a statewide increase of 2.1 percent.[5] Numerical equality required adding approximately 45,000 people to District 2, but not just any 45,000 people if the Delta District was to maintain a black voting-age majority.[6] Concerned that an African American continue to represent the Delta, minority rights advocates wanted to extend its boundary into Jackson, where substantial concentrations of black residents could contribute to a secure 60 percent voting-age supermajority. Their concern was not the electability of Mike Espy, whose conservative roots and family values appealed to white fundamentalist voters. Espy's popularity, black activists feared, made him a strong contender for a statewide office or federal appointment—and likely to be replaced by a white candidate once he gave up his House seat. Philosophically and politically leery of affirmative action, Espy disagreed with black leaders on the need for a supermajority.

Equally resistant to black demands were Espy's white colleagues in the Mississippi delegation, who also valued black voters' loyalty to the Democratic Party. What the Delta District gained, politicians realized, another district lost, and whatever advantage flowed to Espy's hypothetical successor drained the strength and electability of other Mississippi Democrats.

In the end the legislature compromised on a 58 percent black voting-age majority and a 63 percent black majority overall. As figure 3.1 describes, reinforcing the Delta District's minority majority while achieving a court-proof 1.0028 disparity ratio required some intricate political cartography.[7] The new plan splits eleven counties, eight of them on District 2's perimeter. Although I had to generalize district boundaries in places, my map captures the essence of a line that meanders east and west, back and forth, and occasionally up and down to snatch pockets of largely black residents from Districts 1, 3, and 4. Although the new boundaries leave most of the state's eighty-two counties unchanged as well as intact, numerical equality demands that all districts share at least one split county. Least affected is District 5, in the southeast, which grew by 6.2 percent and had population to spare. District 4, which lost black residents in Jackson to the Delta, reached eastward into areas formerly part of Districts 3 and 5. I could go on, but I won't: despite complex borders, Mississippi's new map lacks outrageously bizarre shapes.

If a minority-majority district is to consistently elect a minority candidate, conventional wisdom dictates the need for more than a simple majority. At a minimum, a slight preponderance in voting-age population requires an edge of several percentage points to compensate for the higher birthrates and larger families that make black and Hispanic citizens younger on average than their white counterparts. In Mississippi, for instance, only 63.1 percent of the state's black population was eighteen or older in 1990, in contrast to 71 percent of the total population.[8] At these rates, a minimal 50.1 percent black voting-age majority would require a population 54.5 percent black. But that slight an edge will hardly ensure victory. It's a political fact of life that blacks and Hispanics tend to have lower rates of voter registration and lower voter turnouts than whites.[9] How much lower is uncertain, as is the margin needed to guarantee victory. And much depends, of course, on who is running against whom. Although a popular black officeholder might survive nicely without a majority, a substantial supermajority seems necessary if a black challenger is to unseat a white incumbent.

A rule-of-thumb threshold is 65 percent of the total population. At least that's the number tossed around in the political science literature and various federal court decisions.[10] In 1984 a federal court in Mississippi even broke down the 15 percent population "enhancement" into separate 5 percent adjustments for youthfulness, registration, and turnout.[11] And according to *Fortune,* 65 percent is also the quota-like goal favored by the National Association for the Advancement of Colored People.[12]

Is 65 percent justified? Maybe. Maybe not. Like most guidelines, it's a cautious expedient: high enough to guarantee the election of minority candidates most of the time, yet not likely to waste too many minority votes. And as an obviously rounded multiple of five, it warns of inherent imprecision and uncertainty. But as election consultant Kimball Brace and his colleagues have shown, the highly variable "equalization percentage" required for an effective minority majority—the edge needed for a merely competitive contest rather than a safe seat—is "almost never as high as 65 percent."[13] Moreover, Brace's data demonstrate that the needed margin not only declined during the 1980s but varies enormously from place to place as well as between primary and general elections.

Aware of these deficiencies, the Department of Justice never overtly invoked the 65 percent threshold in its preclearance decisions.[14] Had Justice enforced this quota, Mississippi's reconfigured Delta District would have a more intricate border, and neighboring districts would be less black and more Republican. But 63 percent seems to have worked well enough for both African Americans and Republicans. In April 1993, in a special election following Mike Espy's resignation to become secretary of agriculture, the Delta elected a black successor by a 55 to 45 percent margin.[15] And in 1994 District 1, just east of the Delta, replaced its retiring eighty-four-year-old Democratic incumbent with a conservative Republican, who beat his moderate Democrat opponent by a 63 to 37 percent margin.[16] Although the transfer of black voters to the Delta District hardly accounts for this Republican landslide, a reduced African American presence no doubt contributed to the candidates' decision to run as well as to their campaign rhetoric.

North Carolina in the 1990s was at least faintly akin to Mississippi in the 1980s. Both states have substantial black populations and histories of racial segregation, and both incurred repeated judicial meddling with their congressional boundaries during those decades. Beyond these similarities the analogy weakens. Unlike Mississippi, which is likely to lose a House seat in the year 2000 because of chronic slow growth, North Carolina picked up a seat in the post-1990 reapportionment.[17] With well over twice the population of Mississippi, North Carolina has over one and a half times as many African Americans but lacks the Delta's compact cluster of predominantly black counties. What's more, North Carolina's problems with the courts stem not from calculated resistance but from obeying a Justice Department order to create two minority-majority districts, one of which was a blatant bushmander.

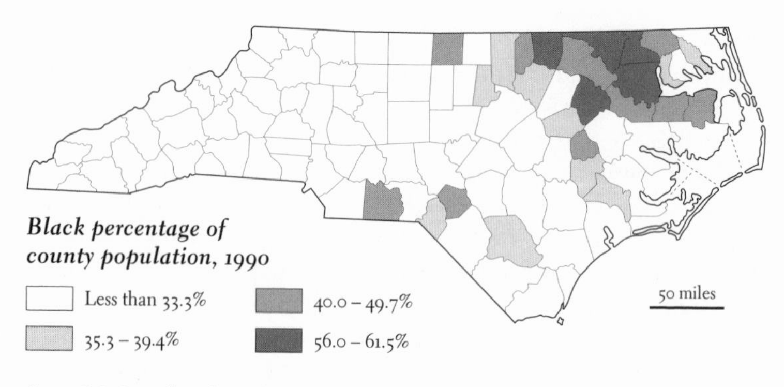

Figure 3.2. Few of North Carolina's one hundred counties are predominantly black.

Where the story begins is uncertain. In the 1980s North Carolina Republicans, a rising force in the state, occasionally joined black leaders in challenging legislative and municipal districts that split minority neighborhoods.[18] Where successful, their lawsuits often resulted in white, pro-Republican suburban districts as well as minority-majority inner city districts. Because adjacent minority communities readily satisfied population thresholds for city council and county supervisor districts, the plaintiffs had little difficulty delineating acceptably compact black-majority districts. By contrast, congressional districts, which require much larger populations, challenged the litigants' ingenuity, knowledge of geography, and willingness to play the race card. Although African Americans made up 22 percent of North Carolina's population, most of them lived in widely separated clusters not readily corralled by straightforward boundaries running between rather than through counties.

In the early 1980s, before the *Gingles* decision, the coalition focused on crafting "minority influence" districts, in which bloc voting might determine the winner's politics if not his (or her) color. In the early 1990s, when the winner's race was the paramount issue, activists vigorously lobbied the Justice Department's newly assertive Voting Rights Division for at least one minority-majority district. An obvious location was the rural northeast quarter of the state, where African Americans made up at least a third of the population in eighteen counties (fig. 3.2). Not as geographically compact or densely black as the Mississippi Delta, the region allowed the legislature to carve out a district with a 51 percent black majority among registered voters: minimal by the NAACP's supermajority standards but sufficient, Democratic leaders hoped, to appease Washington's Section 5 watchdogs.[19]

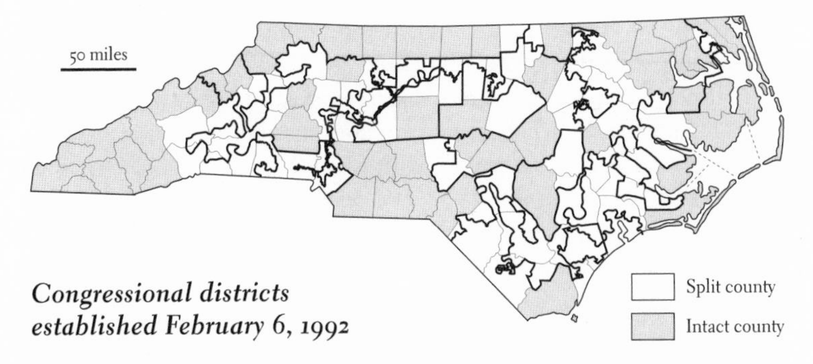

Figure 3.3. North Carolina's post-1990 congressional district boundaries. Compiled from U.S. Bureau of the Census, *Congressional District Atlas: 103rd Congress of the United States* (Washington, D.C., 1993).

Confident of one district, minority activists and their Republican allies pressed for a second minority-majority district, which might (they hoped) dislodge a Democratic incumbent or two. When the legislature's redistricting committee balked and sent Washington a single minority district in July 1991, state Republicans offered an alternative remap, with a second minority district in the state's southeast corner. Although African Americans composed only 48 percent of the district, another 7 percent of residents were Lumbee Indians, considered likely to support a black candidate.[20] Confronted with graphic evidence that a second minority-majority district was possible, the Justice Department's Section 5 unit rejected the overall plan but endorsed the black-majority district in northeastern North Carolina. The price of preclearance, it was clear, was a second district.

With candidates' filing deadlines approaching, legislators fussed, fumed, and ultimately capitulated. On January 24, 1992, North Carolina sent Washington a revised plan (fig. 3.3) with two black-majority districts: a largely rural District 1, with slightly more African Americans than the single black-majority district in the July 1991 submission, and a relatively urban and far less compact District 12 radically different from the mixed black-Lumbee district suggested by Republican strategists. Hesitant to displace a Democratic incumbent in the southeast, the Democratic-controlled redistricting committee focused on the center of the state where Interstate 85 runs west from Durham to Greensboro and then southwest past Winston-Salem to Charlotte. According to the 1990 census, none of these four cities was predominantly black, but their collective total of 307,000 African

Americans was more than sufficient for a minority majority in a district with the roughly 550,000 people required for numerical equality.[21] The challenge, of course, was to connect these cities' predominantly black neighborhoods, pick up additional minority residents in between, and exclude as many white residents as possible—not an easy task if compactness is a concern.

Compactness was apparently less important than survival to Democratic representatives Charlie Rose and Bill Heffner, who would lose loyal constituents under the Republican proposal. According to Capitol Hill scuttlebutt, John Merritt, an aide to Rose, went to work with a computer, digital maps, census data, technical advice from the Democrats' National Committee for an Effective Congress, and moral support from the NAACP.[22] The result was a district the *Washington Post* compared to a "jellyfish tentacle."[23]

Making my small-scale map of North Carolina's post-1990 congressional districts (fig. 3.3) proved a tedious task, at least five times as time consuming as its Mississippi counterpart. The exercise also opened my eyes to the intricate, ambiguously devious twists and turns of district boundaries crafted to win preclearance as well as protect incumbents. I had seen small-scale maps of the North Carolina districts before, in newspapers and magazines as well as in academic journals and law reviews, and had used a similar small-scale map of the I-85 District, compiled from secondary sources, in my chapter on redistricting for *Drawing the Line: Tales of Maps and Cartocontroversy.*[24] But this was the first time I worked directly with maps in the *Congressional District Atlas* for the 103rd Congress, which not only confirmed what I had read (and written) about the much maligned District 12 (the official name of the I-85 District) but exposed similarly contorted lines on the perimeter of District 1 (the black-majority district in the northeast). Although media accounts have focused on the more truly bizarre District 12—its boundary consumes all or part of thirty pages in the *Congressional District Atlas* for the 103rd Congress—District 1's intricate border and thumblike appendages seem equally egregious. What's more, close inspection of the atlas reveals occasionally complex boundaries separating districts not constructed (overtly at least) with race in mind.[25]

What I saw were thin corridors, often less than one hundred feet wide, as in figure 3.4, which shows the border between Districts 9 and 12 meandering across the city of Gastonia to encircle the black neighborhood bounded by Hunt Avenue, North New Hope Road, Separk

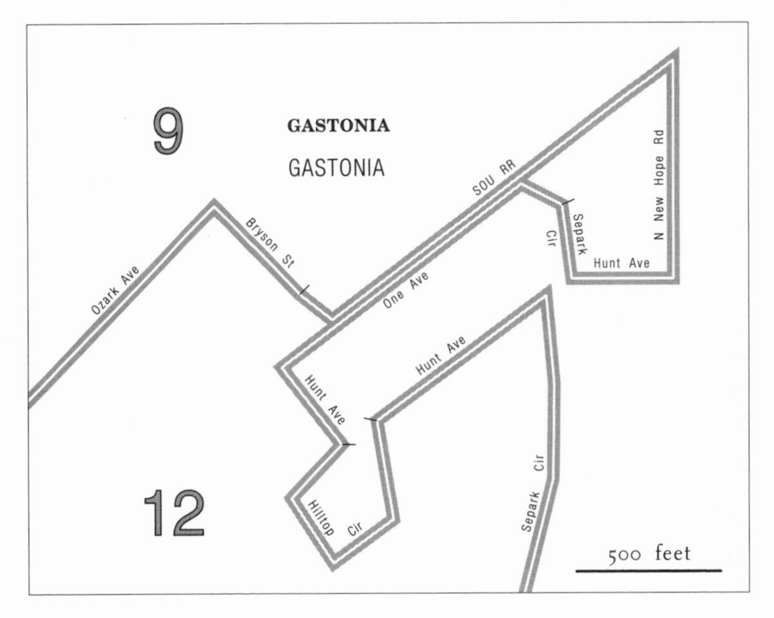

Figure 3.4. Bounded by District 9 along One Avenue and the Southern Railway, District 12 occupies a corridor less than one hundred feet wide. Excerpt from U.S. Bureau of the Census, *Congressional District Atlas: 103rd Congress of the United States* (Washington, D.C., 1993), North Carolina section, 25, inset D.

Circle, and the Southern Railway tracks.[26] Narrow corridors are impossible to portray precisely on a small-scale map like figure 3.3 because the cartographic line symbols, running parallel to each other, not only overlap but greatly exaggerate the corridor's width. Represented on my map by lines 0.01 inch wide—noticeably thicker than county boundaries, because the map's theme requires that district boundaries stand out—each line symbol occupies a path with an equivalent width on the ground of approximately 1.2 miles.[27] To minimize cartographic spaghetti with ugly, meatball-like blobs, I not only smoothed and greatly simplified district boundaries but trimmed off shorter or thinner appendages, as mapmakers commonly do when compiling small-scale maps from highly detailed, large-scale sources. The resulting map (and others like it) captures the districts' overall shape and trend but little of their boundaries' local complexity. Even so, if my drawing conveys at least a general sense of the real map's convoluted borders, it works as well as any attempt to compress fifty-four pages of the *Congressional District Atlas* onto a fraction of a much smaller page.

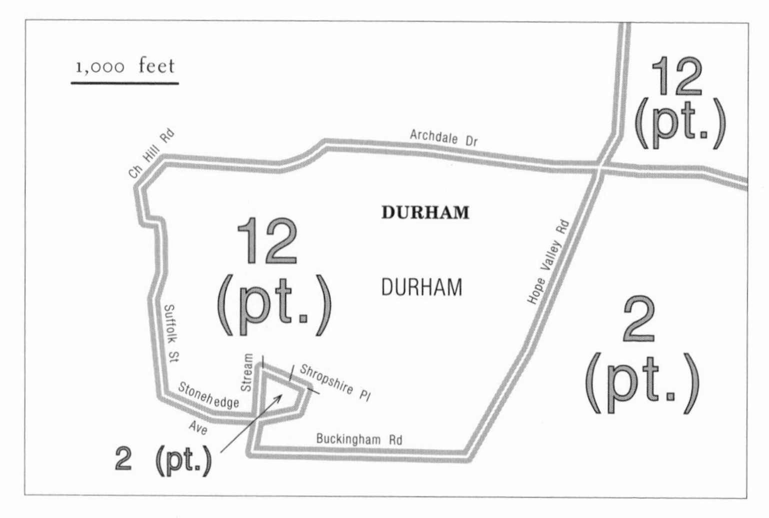

Figure 3.5. Districts 2 and 12 converge to a point twice, so that a part of District 2 is nested within a part of District 12, nested in turn in a part of District 2. Excerpt from U.S. Bureau of the Census, *Congressional District Atlas: 103rd Congress of the United States* (Washington, D.C., 1993), North Carolina section, 18, inset B.

News accounts of North Carolina's remap had prepared me for thin corridors and delicate appendages. After all, it's hard to ignore conservative columnist James J. Kilpatrick's observation that District 12 "rather resembles a lower intestine"[28] or fail to visualize the geography behind candidate Mickey Michaux's promise to hold "campaign rallies at every exit along I-85 from Vance County all the way to Mecklenburg County."[29] But I was not prepared for numerous instances of what geographers call point contiguity: boundary lines that converge to a point and then diverge, as in figure 3.5, where District 2 completely surrounds a portion of District 12.[30] This example from the city of Durham is especially unusual because the map shows a largely black enclave that in turn encloses the largely nonblack portion of District 2 bounded by Shropshire Place, Stonehenge Avenue, and a stream. Technically and legally Districts 2 and 12 remain contiguous, but a person could not walk from the larger part of either district into its outlier without passing over and through its neighbor.

Suppressing enclaves and eliminating short appendages made the map less confusing but didn't allow room for all twelve district numbers, at least not with legible, aesthetically satisfactory labels. So I constructed a second illustration, focusing on districts, not boundaries. To emphasize overall shape, I drew the districts as silhouettes,

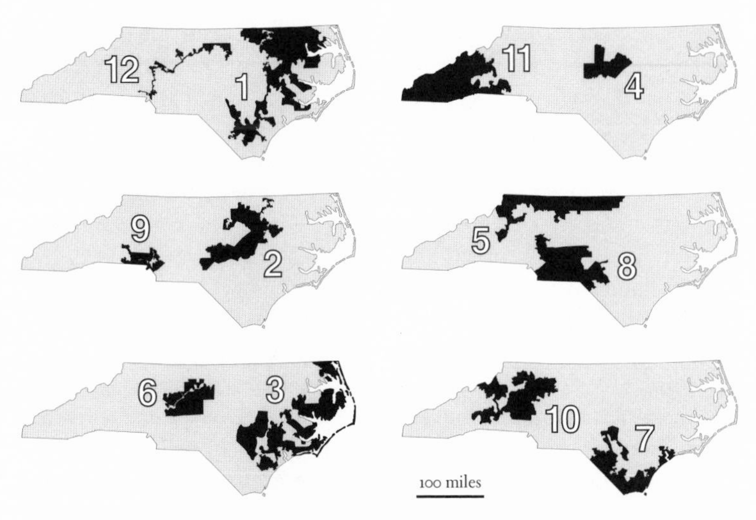

Figure 3.6. Silhouettes of North Carolina's post-1990 congressional districts. Compiled from U.S. Bureau of the Census, *Congressional District Atlas: 103rd Congress of the United States* (Washington, D.C., 1993).

widely separated in groups of two, on small maps with the state outline as a frame of reference (fig. 3.6). An example of what graphics guru Edward Tufte calls "small multiples," this design discloses an abundance of complex shapes, natural as well as artificial.[31] The two minority-majority districts (in the upper left map) differ markedly in size and shape. District 1, the state's largest congressional district, shows the dispersal of rural blacks along the Carolina coastal plain, where many African Americans work in agriculture. By contrast, District 12, the smallest, reflects concentrations of urban blacks on the Piedmont in small and medium-sized industrial cities.[32] However unusual in shape, the I-85 District is more uniform economically and demographically than many of its more compact counterparts.

Districts 1 and 12 are not the only eccentric shapes. District 10 consists of two large portions connected by a thin corridor, and District 6 is actually a trio of similarly substantial sections, separated by District 12 and joined only at points. (Although ink filling the narrow channels cut by the I-85 District hides the separations, a careful comparison of the upper-left and lower-left maps shows that District 12 slices across District 6 from northeast to southwest and makes a second incision to the northwest.) Elsewhere on the display long, gnarled appendages from Districts 2 and 3 cut deeply into District 1 to extract substantial pockets of white voters, and an armlike portion of District 7 nearly encircles a northwest-trending salient of District 1, reaching outward for black voters. Although most of the maps' contorted boundaries reflect the interaction of a minority-majority district with its neighbors, other borders include noteworthy anomalies. Districts 10 and 11, for instance, interlock like pieces of a jigsaw puzzle, as do Districts 5 and 10. But none of the twelve districts is as complex as District 3, situated between District 1 and the drowned valleys and barrier islands of the Carolina coast. Nature, it's clear, can be as geometrically perverse as a political cartographer.

Fascinated with race, the news media looked largely at the state's black-majority districts, which precipitated a torrent of pejoratives and quotable metaphors. Some comments addressed the entire remap, which House Speaker Daniel Blue (who is black) called "an ugly plan,"[33] and Durham lawyer Robinson Everett (who is white) labeled "political apartheid."[34] A few mentioned District 1, which the *Wall Street Journal* branded "computer-generated pornography"[35] and Governor Jim Martin compared to "an upside-down bull moose."[36] But most referred to District 12, which John Dunne, assistant United States attorney general for civil rights, conceded "certainly runs afoul of the principles of compactness"[37] and Republican state chairman Jack Hawke Jr. called "an ideal case for the Supreme Court."[38] *USA Today* labeled the district "a pacesetter in computerized creativity,"[39] while the *Greensboro News and Record* called it "an error" that needed correction but recognized the "radically different concept [that] knitted together a string of urban black neighborhoods."[40] Sympathetic but disappointed, the North Carolina League of Women Voters warned that campaigns would be "extremely expensive" because the district "appears to include at least seven different media markets."[41] Ridicule was rampant. Republican state senator Donald Kincaid suggested that it "looks like we are creating a congressman for motorists,"[42] and Democratic legislator Mickey Michaux joked that "if you drove down the interstate with both doors open, you'd kill most of the people in the district."[43] The *Boston Globe* called it the "ribbon" district,[44] and the *Raleigh News and Observer* compared it to a snake.[45] Despite its awkward shape, the blatant racial gerrymander won over most Democratic officials with its dual promise of preclearance and partisan benefits. "This plan stinks," Democratic congressman John McLaughlin confessed, "but I'm going to hold my nose and vote for it."[46]

Republicans were disgusted and angry. Their outrage reflected not only the atrocious shapes of Districts 1 and 12 but also the skill with which opponents in the legislature had thwarted their hopes of replacing Democratic incumbents with a single black Democrat and a few white Republicans. As University of North Carolina political scientist Thad Beyle observed, party leaders had "outsmarted themselves" by opposing the state's original plan.[47] By contrast, black leaders were delighted with the new districts, neither of which contained a white incumbent. Because African Americans made up 57.3 and 56.6 percent, respectively, of the districts' total populations and 53.4 and 53.3 percent of their voting-age populations, black victories seemed likely if not assured.[48] But minimal majorities left the districts with substantial numbers of wasted white votes, Democratic as well as Republican. Convinced that the legislature and the Justice Department had violated their constitutional right to select a candidate of their choice, some of them sued.

Shaw v. Reno was the defining redistricting case of the 1990s, and the plaintiffs, ironically, were all Democrats.[49] Shaw was Ruth Shaw, a white Durham resident of District 12, and Reno was Janet Reno, Bill Clinton's attorney general and the cabinet officer with oversight responsibility for voting rights enforcement and Section 5 preclearance. Joining Shaw were four other white Durhamites, including Duke University law professor Robinson Everett, who believed "the Constitution is color blind" and adamantly opposed all racial preferences.[50] A former chief judge of the U.S. Court of Military Appeals, Everett was also the group's attorney and chief architect of the lawsuit, filed in federal district court in Raleigh in mid-March 1992 as *Shaw v. Barr.* (George Bush was still president, and William Barr was Bush's attorney general.) The complaint charged that the Justice Department, by forcing the legislature to base boundaries on race, had violated the plaintiffs' Fourteenth Amendment rights to equal protection. In April a three-judge federal panel rejected the claim, and Everett appealed to the Supreme Court, which saw the constitutional significance and agreed, that December, to hear the case.[51] Meanwhile, voters performed as predicted by electing two African American Democrats: Eva Clayton in District 1 and Melvin Watt in District 12. Thanks largely to the Bush administration's Justice Department, North Carolina had its first black congressional representatives in the twentieth century.

After hearing oral arguments in April 1993, the high court announced its decision on June 28.[52] A 5-4 majority of the justices concluded, somewhat inconclusively, that North Carolina's minority-majority districts *might* violate the Constitution. Writing for the majority, Justice Sandra Day O'Connor focused on shape and the perception of fairness: "Reapportionment is one area in which appearances do matter," she argued.[53] Noting that bizarre shapes can affect public perceptions, she quoted a *Wall Street Journal* description of District 1 ("a bug splattered on a windshield")[54] and a graduate assistant's poetic parody of District 12 ("Ask not for whom the line is drawn; it is drawn to avoid thee").[55] Equally important were the shapes' historical implications: "It is unsettling how closely the North Carolina plan resembles the most egregious racial gerrymanders of the past."[56] Despite serious misgivings that "racial classifications with respect to voting...may balkanize us into competing racial factions," the Court declined to rule directly on the appellants' claim. Instead, the justices sent the case back to Raleigh for "close judicial scrutiny."[57] Look at the numbers, the geography, and the Constitution, Justice O'Connor told the lower court, and weigh the impact against "a compelling government interest."[58]

Thirteen months later the district court issued a two-part ruling: Yes, the plan did classify voters by race, but no, it was not unconstitutional.[59] Clayton and Watt were elated, of course, and Everett appealed to the Supreme Court, which would be forced to deal definitively with the constitutionality of contorted racial gerrymanders. Although the lower court had found the North Carolina plan acceptable, the high court's decision in *Shaw v. Reno* had encouraged similar lawsuits in other states, where district shapes were equally or even less suspicious.[60] Ironically, a week before the federal panel in Raleigh approved the North Carolina plan, a three-judge panel in New Orleans condemned a Z-shaped black-majority district in Louisiana said to resemble the "mark of Zorro." And several weeks later a federal panel in Atlanta ruled against a blobish black-majority district in Georgia, which opponents had compared to "Sherman's March."[61] If appearances mattered, as Justice O'Connor contended, what historical or demographic circumstances could validate North Carolina's First and Twelfth Districts yet condemn Louisiana's Fourth District and Georgia's Eleventh Districts, shown in figure 3.7?

So the case went back to Washington, this time as *Shaw v. Hunt.* (James Hunt was the governor of North Carolina; the 1993 decision had taken Janet Reno and the Justice Department out of the case as defendants.) On June 13, 1996, in a decision lawyers and judges call

Figure 3.7. Georgia's Eleventh District and Louisiana's Fourth District are similar in complexity to North Carolina's minority-majority districts. Compiled from the *Congressional District Atlas: 103rd Congress of the United States* (Washington, D.C., 1993).

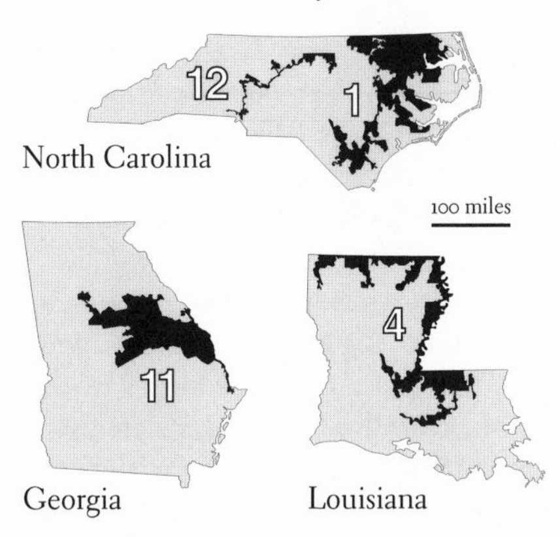

Shaw II, the Supreme Court agreed with Ruth Shaw and her fellow appellants that the state's remap was unconstitutional.[62] In the majority opinion, Chief Justice William Rehnquist summarized the high court's earlier decision as well as the lower court's more recent ruling, which he and his colleagues reversed in a 5-4 vote. His decision recycled words from *Shaw I:* "We now hold that the North Carolina plan does violate the Equal Protection Clause because the State's reapportionment scheme is not narrowly tailored to serve a compelling state interest."[63]

Rehnquist took a less strident view of shape than O'Connor. He conceded that "by anyone's measure, the boundary lines of Districts 1 and 12 are unconventional" but did not dwell on the importance of appearance.[64] Even so, visual impact buttressed his interpretation of District 12 as a unnecessary remedy for a nonexistent violation of the Voting Rights Act. The I-85 District, he argued, "could not remedy any potential §2 violation [because] a plaintiff must show that the minority group is 'geographically compact' to establish §2 liability. No one looking at District 12 could reasonably suggest that the district contains a 'geographically compact' population of any race. Therefore where the district sits, there neither has been a wrong nor can be a remedy."[65]

Justice John Paul Stevens disagreed. In a vigorous rebuttal, he contended that "District 12's noncompact appearance also fails to show that North Carolina engaged in suspect race-based districting." After all, he argued, "there is no federal statutory or constitutional requirement that state electoral boundaries conform to any particular ideal of geographic compactness."[66] And in failing to consider "the political concerns of historically disadvantaged minority groups," the majority opinion was "seriously misguided."[67] More relevant was the overall outcome: "Because I have no hesitation in concluding that North Carolina's decision to adopt a plan in which white voters were in the majority in only 10 of the State's 12 districts did not violate the Equal Protection Clause, I respectfully dissent."[68] Two other justices concurred.

Majorities rule, even slim ones, and the Rehnquist court clearly was not buying highly irregular racial gerrymanders, however remedial and well intended. On the same day that the Court struck down North Carolina's I-85 District, the same five justices ruled against three Texas districts: two with black majorities and one predominantly Hispanic.[69] The North Carolina and Texas decisions were amplifications of concerns expressed two years earlier in *Miller v. Johnson,* when five of the Court's nine justices struck down a black-majority district in Georgia (District 11 in fig. 3.7), largely because "race was the predominant, overriding factor in the decision to attach various appendages containing dense majority-black populations."[70] Applying this "race as the predominant force" standard, federal courts ruled against similarly bizarre minority-majority districts in Florida, Louisiana, New York, and Virginia.[71]

With its congressional districts declared illegal, North Carolina faced the daunting questions of how and when to revise its map. The general election was only five months away: too little time, the state contended, to redraw borders, obtain Justice Department approval, redo the primary election held the previous month, allow candidates adequate time to circulate new nominating petitions and campaign, and deal with litigation that might well send the legislature back to the drawing board yet again.[72] Reshaping the boundaries would be far easier than addressing the inevitable legal challenges: little more than a month after the Supreme Court sentenced Districts 1 and 12 to cartographic reexecution, the legislature was debating a new map with simpler shapes and more muted racial overtones. Although in apparent accord with the Court's decision, the new districts incurred the immediate wrath of black leaders insulted by voting-age populations 51.47 and 51.76 percent *white.*[73] If the state adopted this new map, the NAACP would surely sue. Over the objections of Robinson Everett ("Two elections under an unconstitutional and unfair system is enough"),[74] state officials requested a reprieve. A lower court granted the request, Everett appealed, and in late August the

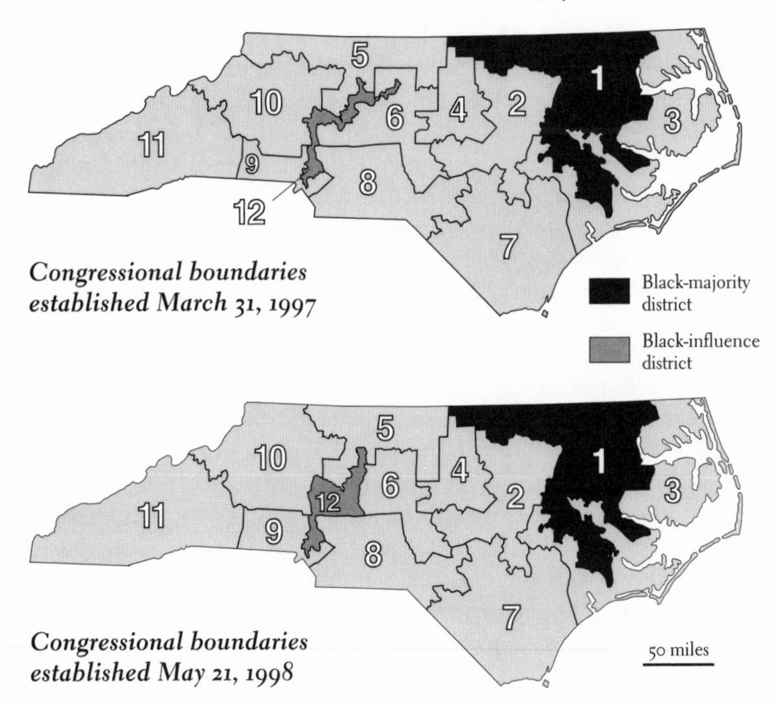

Figure 3.8. In 1998, after the federal district court struck down North Carolina's 1997 redistricting plan (above), the state legislature drew up a somewhat fuller, less irregular version of District 12, which was now only 35 percent African American. Used for the 1998 elections, the new map (below) was abandoned in 1999, when the Supreme Court ruled that the district court had erred. State officials promptly resurrected the 1997 map for the 2000 elections. Compiled from maps on the North Carolina General Assembly Web site.

high court agreed to allow November elections based on May's primary and current boundaries.[75]

Concerned that a postponement not become a pardon, the district court ordered North Carolina to draw up a new map by April 1 of the following year.[76] The legislature's two houses produced separate plans, which leaders sought to resolve as the deadline approached. Fearful that an impasse would invite federal judges to redraw the boundaries themselves, the two houses agreed to make District 1 more compact, cut District 12's length from 160 to 105 miles while making it wider, and protect all twelve incumbents.[77] The joint plan (upper half of figure 3.8) also reduced District 1's black majority to a slim 50.1 percent and made District 12 a minority-influence district,

46 percent black. The Justice Department approved, and on September 15, 1997, a three-judge federal panel consented "on the simple basis that its adequacy...has not been challenged by anyone."[78]

But not for long. Robinson Everett looked at Districts 1 and 12, saw vestiges of their earlier configurations, and sued again. On April 3, 1998, a different three-judge panel, in a 2-1 split decision, agreed with Everett, struck down the new map as excessively favorable to black candidates, and ordered the state to radically redraw District 12 by May 22.[79] Ten days later an adamant Supreme Court rejected the state's request to block the lower court's order. On May 21 the legislature scheduled a new primary for mid-September and submitted a new map (lower half of figure 3.8) based on compactness, not race. This latest District 12, the state maintained, "is regular in shape and makes no 'detours' to pick up African-American residents."[80] And with black residents comprising only 35 percent of the population, it "does not qualify as a minority-majority district by any measure." Although Robinson Everett threatened a further challenge, the Justice Department granted preclearance on June 8, and a three-judge panel approved the new boundaries two weeks later.[81] In endorsing the new map, the judges proclaimed pointedly that race was no longer the predominant factor.

But that was not the last word. On May 17, 1999, the Supreme Court spoke again, this time with one voice. In an unusual unanimous decision, the high court reversed the three-judge panel that a year earlier had ruled against the 1997 remap (upper half of figure 3.8).[82] The lower court had erred in issuing a summary judgment without allowing the state to argue at a trial that its underlying motives were more partisan than racial. Writing for the Court, Justice Clarence Thomas observed that "our prior decisions have made clear that a jurisdiction may engage in constitutional political gerrymandering, even if it so happens that the most loyal Democrats happen to be black Democrats and even if the state were *conscious* of that fact."[83] What's more, "Evidence that blacks constitute even a supermajority in one congressional district while amounting to less than a plurality in a neighboring district will not, by itself, suffice to prove that a jurisdiction was motivated by race in drawing its district lines when the evidence also shows a high correlation between race and party preference."[84] With the 1998 remap no longer necessary, state officials resurrected District 12's markedly less regular 1997 boundaries for the next election.[85]

Figure 3.9. Illinois's Fourth District and New York's Twelfth District have Hispanic majorities, but only the New York district was judged unconstitutional. Compiled from the *Congressional District Atlas: 103rd Congress of the United States* (Washington, D.C., 1993).

North Carolina might easily feel abused—an innocent victim caught between an aggressively proactive Justice Department and an adamantly resistant yet numerically minimal Supreme Court majority. In this sense any state compelled to create and then dismantle a minority-majority district is also a victim of fate. After all, the outcome might have been radically different had the death or retirement of one of the five "color-blind" justices afforded Bill Clinton another appointment early in his presidency.

For readers who disdain iffy history, there's the luck of Illinois, which in the early 1990s created an earmuff-shaped Hispanic-majority district (fig. 3.9) little different in intent and complexity from New York's blatantly unconstitutional Bullwinkle District. Why District 4 in Illinois escaped the fate of District 12 in New York or District 12 in North Carolina puzzled many politicians. In March 1996 a three-judge federal panel in Chicago decided that the district "passes constitutional muster" despite its noncompact shape.[86] "Where the drawing even of irregular lines is required to remedy established violations of the Voting Rights Act, the court need not flinch from its obligation to do so with a bold and deliberate pen."[87] That November, after the plaintiff appealed, the Supreme Court asked the Chicago jurists to reopen the case in the light of recent rulings on racially gerrymandered districts in North Carolina and Texas. When the lower court affirmed its earlier ruling, the plaintiff appealed once again. In

January 1998, with no explanation, the high court declined to hear the appeal and thereby endorsed the lower court's view that the district did not violate the Equal Protection clause.[88]

District 4's boundaries survived, I think, because of savvy political cartographers: when Illinois was unable to deliver a timely remap following the 1990 census, a panel of federal judges stepped in to draw the lines.[89] The judges recognized a "compelling state interest" when they saw one, knew how to craft a "narrowly tailored plan" that did not abuse the Court's discretion, and prepared for the inevitable legal challenge by carefully documenting their rationale. Knowing who would be judging their work, they also knew what would raise eyebrows.

4 *Redrawing the Lines*

WHAT CAN WE LEARN FROM JUDICIAL scrutiny of congressional boundaries? Quite a bit when we compare before and after views for North Carolina, Georgia, Louisiana, and New York, where remaps in which minority composition was not paramount yielded markedly more compact districts. Somewhat less if we also look at states like Illinois, where the courts endorsed an awkwardly shaped district held to address a compelling government interest, or Texas, where the judiciary condemned highly irregular minority-majority districts but largely ignored equally contorted white-majority districts. In addition to probing court-imposed map revisions for insight and meaning, this chapter examines the impact of boundary changes on neighboring districts and the ef-

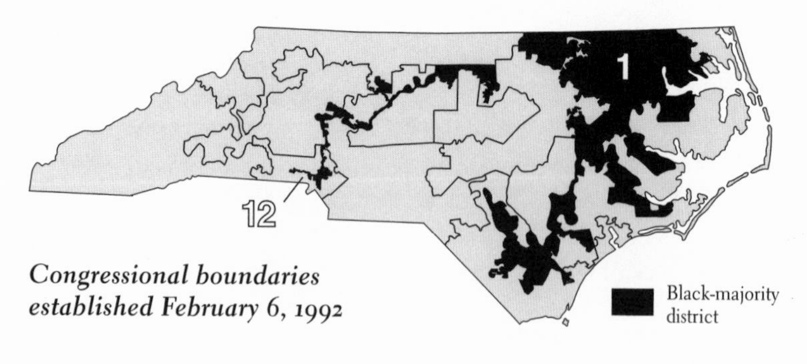

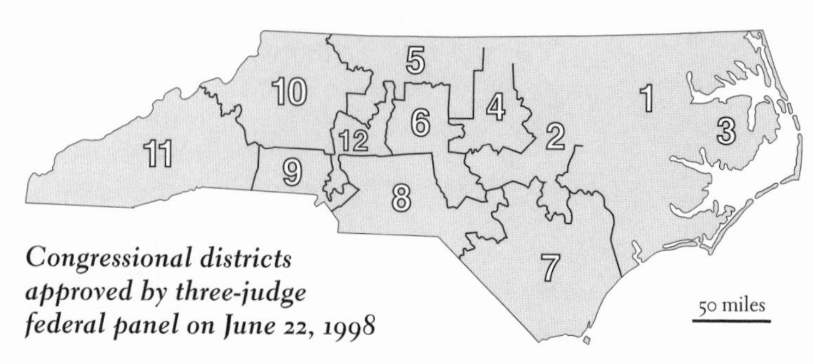

Figure 4.1. North Carolina congressional districts for the 1992 and 1998 elections. Compiled from the *Congressional District Atlas* and an updated map from the North Carolina General Assembly. Because a 1999 Supreme Court decision overturned the lower-court ruling the lower map was designed to address, these boundaries were used in 1998 but not in 2000.

fects of racially motivated districting on minority and Republican candidates.

North Carolina's final remap, approved just in time for the 1998 congressional elections, demonstrates how effectively lowering the minority percentage promotes compactness in states with dispersed minority populations. On the new map (lower half of fig. 4.1), District 12 carried hardly a trace of its earlier association with Interstate 85. No longer reaching east to Greensboro and Durham or west to Gastonia, the new district was only 85 (rather than 160) miles between extreme points. It was also noticeably wider, the result of closer conformity to county boundaries. And rather than encompassing minor parts of ten counties, District 12 now consisted of one complete county and parts of four others. In cutting out Durham, the legislature's political cartographers conveniently warded off further challenges from Robinson Everett, Ruth Shaw, and their fellow plaintiffs. Only bona fide residents, the courts have ruled, can challenge a district's constitutionality under the Equal Protection clause. According to political scientist Charles Bullock, redistricting officials frequently use this ploy to derail further litigation.[1]

Comparing the 1992 and 1998 maps suggests that the legislature was taking no chances with new challenges. Although the Supreme Court's ruling in *Shaw II* ignored the constitutionality of North Carolina's other minority-majority district, the revamped map shows District 1 with a smoother, far less irregular perimeter enclosing a total population (according to 1990 census counts) only 50.3 percent black, down from 57.3 percent under the 1992 plan.[2] Chastened by the Court for its aggressive pursuit of minority majorities, the Justice Department's Voting Rights Section was in no mood to deny preclearance, despite a drop in the black share of the district's voting age population from 53.4 to 46.5 percent.

Nor were legislators taking chances with population equality. Total populations for the twelve new districts—no district survived the remap intact—ranged from 551,842 to 553,143, for an impressively scrupulous disparity ratio of 1.0024. More remarkable is the new districts' comparative compactness. Although the 1992 plan carved up forty-four of the state's one hundred counties, the 1998 map reflects only twenty-one split counties. So open was the new map that I had little difficulty identifying districts by number.

Had a simpler map been its only goal, the legislature would have split far fewer counties. But as practical politicians, redistricting officials were wary of disrupting the newly established partisan balance—a breach of etiquette likely to precipitate a deadlock or lawsuit and goad an agitated trio of federal judges into revising the map themselves. A brief filed with the new map noted that "maintaining District 12 as a Democratic district and neighboring districts 5, 6, 9 and 10 as Republican districts was essential."[3] A reconstituted District 12 with two or three intact counties would have been markedly more Republican, the state's lawyers argued, and the city of Charlotte had to be split because the House members representing Districts 9 and 12 lived there. Incumbents almost always survive minor cartographic surgery, and putting two of them in the same district was not an option.

Although loss of their black voting-age majorities was troubling, North Carolina's two minority incumbents had little cause for despair. Two years earlier a court-imposed remap of Georgia's congres-

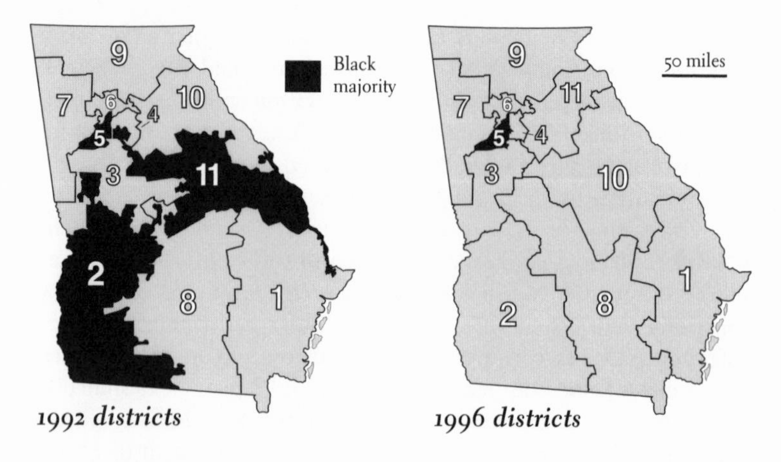

Figure 4.2. Georgia congressional districts for the 1992 and 1996 elections. Compiled from the *Congressional District Atlas* and the updated Congressional Districts State Map Data Sets, 105th Congress, from the Geography Division, Bureau of the Census.

sional districts had cut the number of predominantly black constituencies from three to one (fig. 4.2). Least affected was District 5, in urban Atlanta, where the black share of the voting-age population dropped from 57.5 to 57.2 percent.[4] The relatively rural southwestern corner of the state witnessed a more marked restructuring that reduced the number of African Americans in District 2 from 52 to 35 percent of the population eighteen and older. Although North Carolina's I-85 District experienced a greater dilution of its black majority, from 53 to 33 percent of the voting-age population, neither cutback was comparable to the cartographic amputations and dislocations of Georgia's District 11, the focus of plaintiffs' wrath in *Miller v. Johnson.* The new map not only slashed the black proportion of potential voters from 60 to 11 percent but vaulted District 11 northward, across District 10, where the African American share rose from 16 to 34 percent. In spite of this geographic reshuffling, all three black incumbents won reelection.

So radical was the Georgia remap that several incumbents, white as well as black, moved to districts where they had stronger name recognition and proven voter support. The most prominent political refugee was two-term Democratic congresswoman Cynthia McKinney, the first black woman elected to Congress from Georgia. For the 1996 election, McKinney jumped from old District 11 to new District 4, in the Atlanta suburbs. Although only 33 percent of potential voters in her new district were black, McKinney defeated three white male challengers in the Democratic primary and captured 58 percent of the vote in the general election.[5] Relocation also worked for black congressman Sanford Bishop, whose home landed in District 3 on the new map, along with a white Republican incumbent and a voting-age constituency less than 23 percent black. Bishop wisely moved to District 2, which contained twenty-eight of the thirty-five counties in his old district and had a 35 percent black voting-age population. After defeating two white challengers in the Democratic primary, Bishop won 54 percent of the final vote. The court-imposed remap also displaced two Republican incumbents, who chose to follow their old constituencies. Political migration is not too forbidding under Georgia law, which lets candidates run outside their home districts.[6]

That all eleven of Georgia's House members won reelection in 1996 is hardly surprising: once elected, incumbents enjoy an enormous edge, jeopardized only by a political sea change or drastic redistricting. What put most of them in office was a combination of the post-1990 remap and the Republican party's new, pro–Christian fundamentalist southern strategy.[7] Before the 1992 elections, the state's smaller, ten-person delegation included only one Republican and one black Democrat. In 1992 two new minority-majority districts not only paved the way for Bishop's and McKinney's initial victories but helped Republicans capture three districts where the cartographic weeding out of black voters removed a traditional Democratic advantage. GOP candidates might have fared even better had Ross Perot not undermined George Bush's reelection campaign. In 1994 Republican candidates touting the Contract with America won three more seats, and in April 1995 the remaining white Democrat defected to the GOP.[8] Although the Court-drawn 1996 map added black voters to all but two Republican congressional districts, the Georgia delegation held steady at three black Democrats and eight white Republicans.

No longer constrained by race, the new districts are noticeably more compact (fig. 4.2). Gone are the tentacles with which District 11 reached north and south along the South Carolina border toward black concentrations in Augusta and Savannah as well as westward into Atlanta's black suburbs. Gone too are the bulbous protrusions with which District 2 and its neighbors interlocked like pieces of a giant jigsaw puzzle. Although most of the new districts are noticeably elongated, their borders largely follow county boundaries. In contrast to the 26 split counties on the 1992 map, district boundaries now cut into only 6 of Georgia's 159 counties. All split counties are in the Atlanta area, where District 5, the remaining black-majority constituency, is little changed from its 1992 configuration. Ironically, District 5 is among the state's most compact districts.

Louisiana's experience with minority-majority districts, the Department of Justice, and the federal courts is similar in many ways to Georgia's. Although demographic patterns supported a single compact black-majority district, largely within New Orleans, the state's largest city, Justice officials insisted on a second minority-majority district, which cut across the 1992 map, as pundits gleefully pointed out, like the mark of Zorro (fig. 3.7, lower right). And like Georgia's old District 11, Louisiana's District 4 triggered lawsuits, impassioned pleas to federal judges, a decisive finding of unconstitutional racial bias, and a court-drawn 1996 map that was unquestionably simpler overall in geography and geometry.

Despite these similarities, Louisiana's brush with bushmanders was far more complex than Georgia's. Not only was District 4's twisted, strung-out shape more outrageous than the appendages of its Peach State counterpart, but its eradication involved an intermediate 1994 map (fig. 4.3, center), also with two minority-majority districts, which the courts condemned with equal if not greater vigor. And unlike Georgia's 1996 remap, which all black incumbents managed to survive, Louisiana's ultimate remap proved politically fatal to a black congressman forced to share a new white-majority district with a white Republican incumbent.

House member Cleo Fields, the undisputed beneficiary of District 4's 1992 incarnation, handily defeated his first white opponent. With African Americans composing 66 percent of the total population, Fields's constituency ranked eighth in racial makeup on *Congressional Quarterly*'s list of thirty-two black-majority districts.[9] Its racial basis, which the legislature made no effort to hide (How could they?), angered a quartet of Louisiana voters, three white and one black, who challenged the map in federal court. In December 1993 a three-judge panel, informed by the Supreme Court's majority opinion in *Shaw I*, declared the 1992 map unconstitutional and ordered the state to try again. In April 1994 the legislature approved a new map (fig. 4.3, middle) on which a drastically reconfigured District 4 ran southeastward along the Red River from the Texas border to Baton Rouge.[10] Crafted with Congressman Fields in mind, the new constituency was 58 percent black. Although the Justice Department approved the reconfiguration in June, the federal panel repeated its objection to a sec-

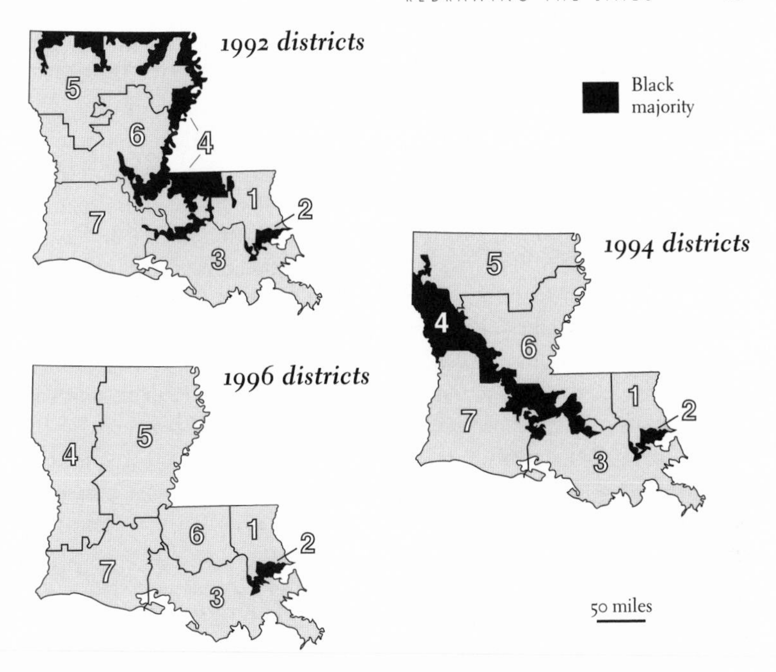

Figure 4.3. Louisiana congressional districts for the 1992, 1994, and 1996 elections. Compiled from the *Congressional District Atlas* and the updated Congressional Districts State Map Data Sets, 104th and 105th Congresses, from the Geography Division, Bureau of the Census.

ond minority-majority district, however more compact, and in late July it imposed its own map, with even simpler boundaries and a single black-majority district.[11] With fall elections and filing deadlines approaching, the state and the Justice Department appealed to the Supreme Court, which in mid-August granted a stay allowing use of the legislature's new map.[12] Discouraged by cartographic uncertainty, few credible challengers emerged, and all seven incumbents, including Congressman Fields, won reelection.[13]

Nearly a year passed before the high court formally addressed the appeal. Its ruling, announced on June 29, 1995, was hardly decisive: because none of the plaintiffs actually lived in District 4, their Equal Protection claims were groundless, leaving the lower court no reason to redraw the map.[14] Although disappointed, the plaintiffs quickly recruited eight residentially plausible victims of District 4 and optimistically refiled their lawsuit.[15] Their confidence reflected the Supreme Court's decision in another redistricting case, *Miller v. Johnson*, announced the same day. In decisively overturning Georgia's Eleventh

District, the five justices who two years earlier had merely questioned the constitutionality of North Carolina's I-85 District and similarly bizarre shapes now decisively denounced race-based districting.

The majority opinion, written by Justice Anthony Kennedy, broadened the Court's ruling in *Shaw I.* "Shape is relevant not because bizarreness is a necessary element of the constitutional wrong or a threshold requirement of proof, but because it may be persuasive evidence that race of its own sake...was the legislature's dominant and controlling rationale in drawing its district lines."[16] Georgia—and by implication Louisiana—had to redraw its map because the lines did not reflect "traditional race-neutral districting principles, including but not limited to compactness, contiguity, [and] respect for political subdivisions or communities defined by actual shared interests."[17] In noting the Court's pointed rejection of Washington's efforts, under both Bush and Clinton, to maximize the number of minority-majority districts, the *Congressional Quarterly Almanac* aptly labeled the *Miller* ruling "a slap at the Justice Department."[18]

Neither the Justice Department nor the Louisiana legislature could guarantee Cleo Fields a safe seat. In early 1996, after a federal judge declared the 1994 map unconstitutional, legislators presented a new plan, indistinguishable from the district court's 1994 map.[19] Boundaries were simpler (fig. 4.3, bottom), there was only one black-majority constituency, and Fields landed in a district with a white Republican incumbent. The Clinton administration, the Louisiana Legislative Black Caucus, and the state's attorney general appealed to the Supreme Court to keep the second black district, but in late June the justices upheld the district court in a decisive 8-1 ruling.[20] Intimidated by his new district's 71 percent white population, Fields decided not to file for reelection.[21]

Against the backdrop of growing judicial disdain for intricate racial gerrymanders, there was little surprise in the Supreme Court's 1996 rejection of three minority-majority districts in Texas. By a 5-4 majority, the high court upheld a district court ruling that the districts were based predominantly on race and hence unconstitutional. Justice Sandra Day O'Connor, who wrote the decision, rejected claims that the contorted shape of District 30, a black-majority constituency in Dallas, might be "explained by efforts to unite communities of interest, as manifested by the district's consistently urban character and its shared media sources and major transportation lines."[22] And she energetically condemned Districts 18 and 29, Houston constituencies with black and Hispanic majorities, respectively. "Not only are the shapes of the districts bizarre, [but] their utter disregard of city limits, local election precincts, and voter tabulation district lines...caused a severe disruption of traditional forms of political activity [and] created administrative headaches for local election officials."[23] To underscore her points by letting the boundaries speak for their own outrageousness, Justice O'Connor appended silhouettes of the three offending constituencies (fig. 4.4).

What is remarkable about *Bush v. Vera* is not the irony that Republican governor George W. Bush, son of the former president, appealed a lower-court ruling that struck down three bushmanders: like his dad and Elbridge Gerry, George W. was more a bystander than an advocate. What's truly remarkable is that the courts, as if to reinforce the majority view that race, not shape, was at the heart of the matter, failed to condemn several equally contorted and no less dysfunctional white-majority districts. In a spirited, mildly sarcastic dissent, Justice John Paul Stevens not only appended silhouettes of three of them (fig. 4.5) but used a footnote to parody verbal descriptions written by

Minority-majority districts ruled unconstitutional

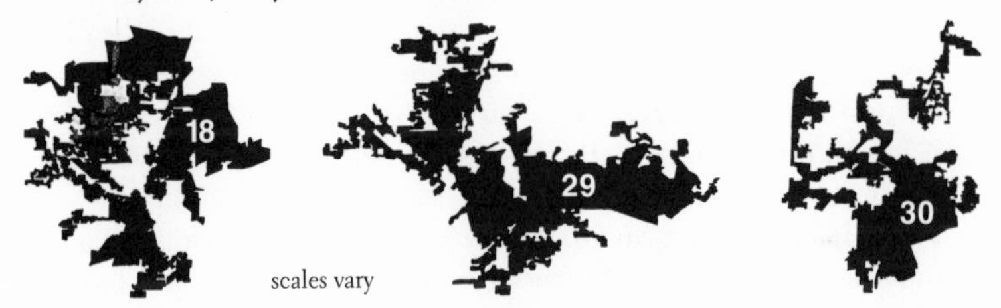

Figure 4.4. Texas minority-majority districts deemed unconstitutional in *Bush v. Vera*, as described in appendixes to Justice O'Connor's plurality opinion.

White-majority districts left unchallenged

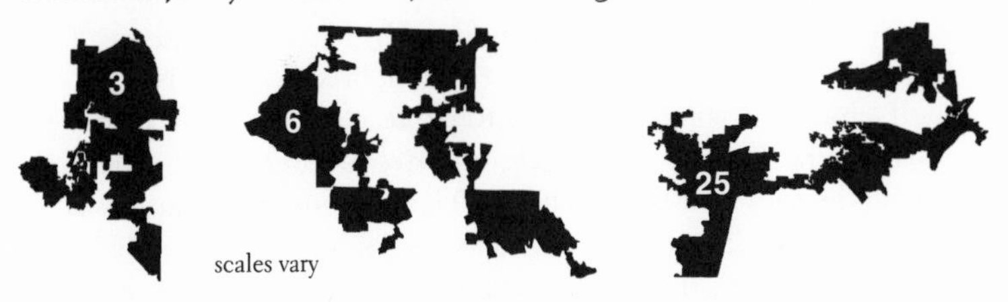

Figure 4.5. Texas white-majority districts tacitly approved by a majority of the justices in *Bush v. Vera*, as described in appendixes to Justice Stevens's minority opinion.

the district court and repeated by Justice O'Connor. His footnote refers to District 6 (fig. 4.5, center), a purely partisan gerrymander with an Anglo majority and a Republican incumbent:

> As for the obligatory florid description: District 6 has far less of an identifiable core than any of the majority-minority districts struck down by the District Court. To the extent that it "begins" anywhere, it is probably near the home of incumbent Rep. Barton in Ennis, located almost 40 miles southwest of Downtown Dallas. From there, the district winds across predominantly rural sections of Ellis County, finally crossing into Tarrant County, the home of Fort Worth. It skips across two arms of Joe Pool Lake, noses its way into Dallas County, and then travels through predominantly Republican suburbs of Fort Worth. Nearing the central city, the borders dart into the downtown area, then retreat to curl around the city's northern edge, picking up the airport and growing suburbs north of town. Worn from its travels into the far northwestern corner of the county (almost 70 miles, as the crow flies, from Ennis), the district lines plunge south into Eagle Mountain Lake, traveling along the waterline for miles, with occasional detours to collect voters that have built homes along its shores. Refreshed, the district rediscovers its roots in rural Parker County, then flows back toward Fort Worth from the southwest for another bite at Republican voters near the heart of that city. As it does so, the district narrows in places to not much more than a football field in width. Finally, it heads back into the rural regions of its fifth county—Johnson—where it finally exhausts itself only 50 miles from its origin, but hundreds of [quoting the majority opinion in *Miller v. Johnson*] "miles apart in distance and worlds apart in culture."[24]

More telling is Justice Stevens's comparison (this time *not* a footnote) of District 6 with a black-majority district five of his fellow jurists deemed illegal:

> For every geographic atrocity committed by District 30, District 6 commits its own and more. District 30 split precincts to gerrymander Democratic voters out of the Republican precincts; District 6 did the same. District 30 travels down a river bed; District 6 follows the boundaries of a lake. District 30 combines various unrelated communities within Dallas and its suburbs; District 6 combines rural, urban, and suburban communities. District 30 sends tentacles nearly 20 miles out from its core; District 6 is a tentacle, hundreds of miles long (as the candidate walks), and it has no core.[25]

In addition to arguing that "the existence of the equally bizarre majority-white District 6 makes the plurality's discussion of District 30's odd shape largely irrelevant," he also suggested that "the incumbency considerations that led to the mutation of District 6 were the same considerations that forced District 30 to twist and turn its way through North Dallas."[26] Is race alone sufficient reason to bless one and damn the other?

Well, yes, conceded Justice O'Connor, who addressed her colleague's complaints in her own opinion. (Before the Supreme Court releases a decision, the justices share and modify their written statements.) "District 30's combination of a bizarre, noncompact shape and overwhelming evidence that shape was essentially dictated by racial considerations of one form or another is exceptional; Texas Congressional District 6, for example, which Justice Stevens discusses in detail...has only the former characteristic. That combination of characteristics leads us to conclude that District 30 is subject to strict scrutiny."[27]

Among civil rights advocates, "strict scrutiny" has a familiar and ominous ring. Increasingly common in decisions involving affirmative action and racial classification, the term describes the close, highly critical reading of evidence by jurists committed to absolute racial neutrality: strict constructionists who see little room under the Fourteenth Amendment for racial quotas, quasi quotas, and purposefully contorted minority-majority districts.[28] In applying strict scrutiny to election districts, conservative judges insist that any apparent racial preference be "narrowly constructed" and serve a "compelling state interest" rather than merely maximize the number of minority candidates likely to be elected.[29] The three Texas districts in figure 4.4 obviously failed this test and had to be redrawn. By contrast, a similarly complex Hispanic-majority district in Illinois (fig. 3.9) passed constitutional muster largely because the lower-court judges responsible for its boundaries had done their homework. At least most of the high court thought so. In affirming the federal panel's wisdom by declining to hear an appeal, the Supreme Court was not unanimous: Justices Antonin Scalia, Anthony Kennedy, and Clarence Thomas, strictly scrutinous to the end, voted to hold a trial.[30]

It had to happen: a showdown between the Justice Department's preclearance mavens and the five Supreme Court justices committed to striking down districts crafted principally to send more blacks and Hispanics to Congress. The high court held the upper hand but was reluctant to come out swinging. Capable of a slap in the face, as in *Miller v. Johnson,* the justices mostly shook a metaphorical fist at the Voting Rights Section, which resisted efforts to simplify boundaries. In August 1996, for instance, the Justice Department withheld preclearance from the Louisiana legislature's court-directed remap: a symbolic gesture of defiance with no practical effect because the backup map was the Court-drawn plan that the frustrated legislators had merely mimicked.[31] And in late 1996 federal lawyers joined civil rights activists in an unsuccessful challenge to the Court-crafted boundaries used in Georgia's 1996 elections. Plaintiffs charged that the Court's remedial map, in wiping out two of the state's three black-majority districts, went well beyond the original *Miller* decision, which struck down only District 11. By a 5-4 majority (What else?) the high court upheld the lower court's map and rejected arguments that the district judges should have sought preclearance.[32]

The showdown came in May 1997, in *Reno v. Bossier Parish School Board,* a case involving school board voting districts in a parish (county) in northwest Louisiana.[33] In 1993 the Justice Department had denied preclearance to a remap that retained white majorities in all twelve districts. In the eyes of department officials, because the parish could have drawn two black-majority districts, the new map failed to satisfy Section 2 of the Voting Rights Act, which outlaws voting practices that limit the opportunity for minority voters to elect candidates of their choice. At issue before the Court was the denial of preclearance under Section 5, which contains the preclearance requirement and applies only to areas with a history of discrimination. Recall that Section 5 insists that changes not dilute minority voting strength. Although the wording of Section 5 appears only to prohibit retrogressive changes, a department regulation issued in 1987 denied preclearance if a different redistricting plan could further enhance the collective clout of minority voters. Using this regulation, the Justice Department pressured Georgia, Louisiana, and other states, largely in the South, into drawing complex congressional, legislative, and municipal-election districts. On May 12, after half a decade of aggressive strict scrutiny, the Supreme Court struck down the 1987 regulation in an uncharacteristically cohesive 7-2 vote: no longer was suspicion of a Section 2 violation grounds for denying preclearance under Section 5.

There's a cartographic lesson in strict scrutiny, which almost always produced smoother, simpler boundaries. In July 1997, for instance,

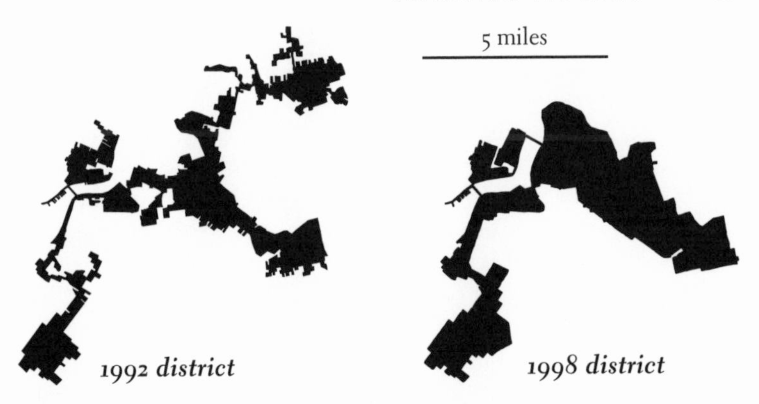

Figure 4.6. New York's District 12, crafted as the state's second Hispanic-majority district in 1992, had smoother boundaries and a less complex shape for the 1998 elections. Right-hand silhouette compiled from map provided by the New York State Legislative Task Force on Demographic Research and Reapportionment.

a court-ordered remap of New York's infamous Twelfth District clipped Bullwinkle's ethnically inspired antlers (fig. 4.6).[34] And seven months later, in the decade's last court-ordered congressional remap, the new boundaries that eliminated Virginia's single black-majority district split two fewer counties and five fewer cities.[35] For earlier court-mandated remaps, similar simplifications had a noticeable effect on the *Congressional District Atlas,* for which the Bureau of the Census published updated "state map sets" for the 104th and 105th Congresses. Between the 1992 and 1996 elections, the number of map pages dropped from 104 to 85 for Florida, from 35 to 8 for Georgia, and from 177 to 96 for Texas. For Louisiana, where successive revisions reduced the number of maps pages from 28 in 1992 to 23 in 1994 and to a mere 8 in 1996, figure 4.3 verifies that the more recent redrafting produced the greater simplification. Despite these adjustments, congressional boundaries were more complex overall at the end of the decade than in 1990.

5 *Gauging Compactness*

IT'S TEMPTING TO SUGGEST A FORMAL standard for compactness: a definition separating the illegally irregular from the geographically gauche. But as the preceding chapters illustrate, compactness is a bit like pornography—although we know it when we see it, individual sensitivities and community standards vary widely. Experience suggests that the best we can do is to measure diverse aspects of district shape and size and adopt legal limits similar to the caps placed on campaign contributions. Yet for various reasons our best isn't good enough. Because transportation systems and natural barriers distort space, ten miles here need not equate to ten miles there. And because rivers, shorelines, ridge tops, and other highly functional political borders meander freely, a contorted boundary is not

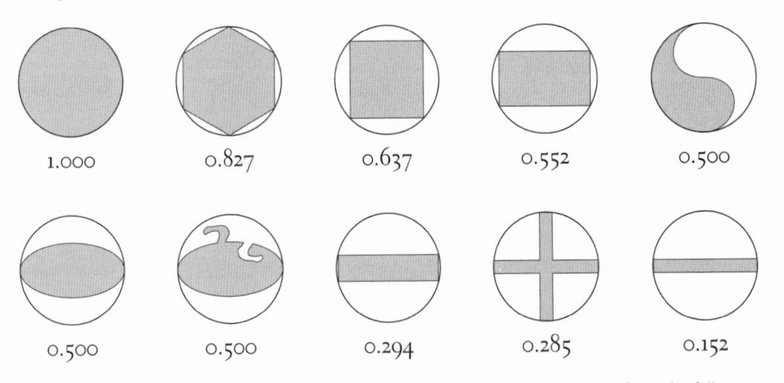

Figure 5.1. Dispersion scores for a variety of shapes illustrate the measurement of circular fullness.

inherently reprehensible. What's more, radical variations in population density undermine the significance of silhouette maps, and islands demand a flexible contiguity requirement. In addition to examining how geographers and political scientists measure shape and differentiate boundaries, this chapter questions the relevance of geometric distance in the age of the Internet and the interstate highway.

Compactness is easy to measure—perhaps too easy. Social scientists have devised more than two dozen indexes to describe the irregularity of political districts and service areas, but comparative studies suggest that two or three basic measures are usually sufficient.[1] Although some shape indexes are highly complex, simple formulas based on straightforward geometric terms like circle, perimeter, and area can capture the fundamental dimensions of a closed geographic boundary.

I focus here on the two indexes with which legal scholars Richard Pildes and Richard Niemi evaluated the appearance of post-1990 congressional districts.[2] Straightforward and complementary, these indexes are dimensionless numbers, unaffected by district size, that range from 1.0 for the perfectly compact to nearly 0.0 for the maximally bizarre.[3]

The first index is the *dispersion score,* computed by dividing the area of a district by the area of the smallest circle that completely encloses the district. As the examples in figure 5.1 attest, using a circle as the standard for compactness affords meaningful distinctions between compact and elongated shapes as well as between thin and full shapes. Note, though, that the dispersion score is not sensitive to incisions or

Perimeter scores

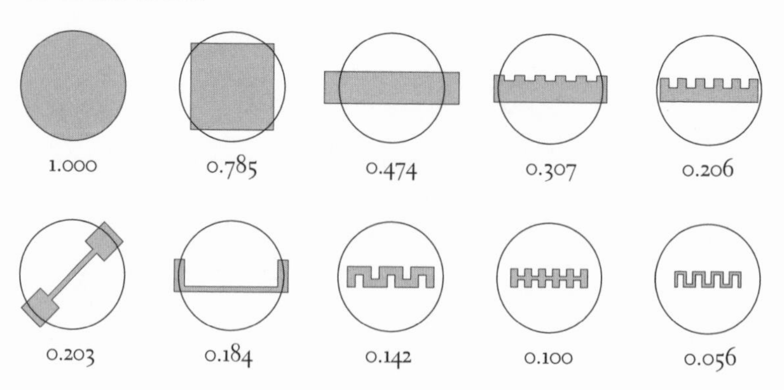

Figure 5.2. Perimeter scores for a variety of shapes illustrate the measurement of boundary efficiency.

appendages that fit within a district's circumscribed circle. Three of the ten examples have identical dispersion scores (0.5), but the ellipse (lower row, second from the left) with a hook-shaped incursion reattached as an appendage—an exchange that lengthens the perimeter but doesn't alter the area—is clearly less regular and arguably less compact than either its unaltered elliptical parent or the equally dispersed yin-yang symbol.

Irregular edges demand a second index, the *perimeter score,* computed by dividing a district's area by the area of a circle with a circumference equal in length to the district's perimeter. By this standard, a perfectly circular district registering 1.0 contrasts markedly with long, thin shapes scoring close to zero. As the hypothetical shapes in figure 5.2 demonstrate, the perimeter score measures the efficiency of district boundaries: think of this index as the relative expense of buying fence by the mile or paying cartographic drafters by the inch. As their identical circles suggest, the ten diverse shapes have identical perimeters. But as the perimeter scores indicate, their boundaries enclose markedly different areas. In general, intricate, irregular edges lower the perimeter score, as do deep, narrow incisions or appendages like those of the mutilated ellipse in figure 5.1. In addressing these local anomalies, the perimeter score captures an element of shape distinctly different from a boundary's overall circular fullness as measured by the dispersion score.

To assess how well these indexes gauge the compactness of real congressional districts, I arranged generalized silhouettes of North

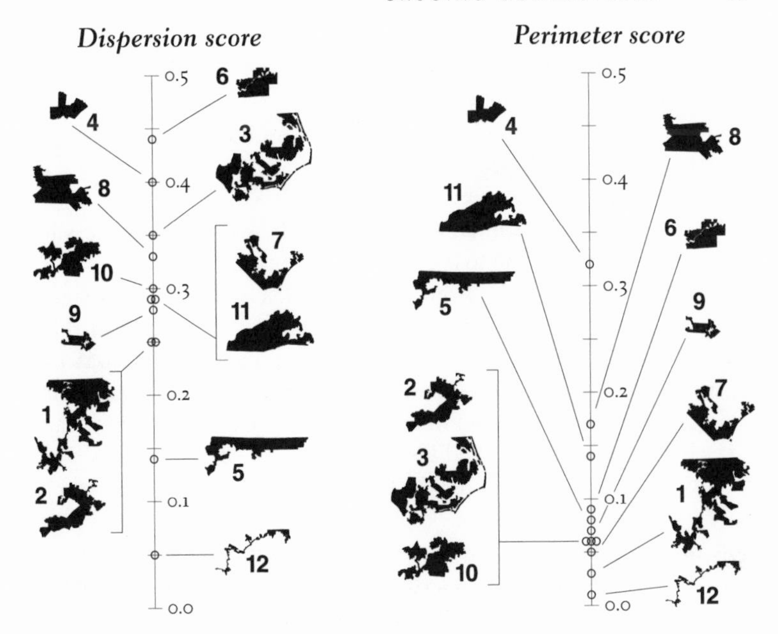

Figure 5.3. Compactness indexes for North Carolina's 1992 congressional districts. Based on dispersion and perimeter scores reported in Richard H. Pildes and Richard G. Niemi, "Expressive Harms, 'Bizarre Districts,' and Voting Rights: Evaluating Election-District Appearances after *Shaw v. Reno*," *Michigan Law Review* 92 (1993): 562, table 2.

Carolina's twelve congressional districts along separate dispersion and perimeter scales. Even though the scores reported in Pildes and Niemi's study are based on more geometrically exact representations of district outlines, the icons in figure 5.3 demonstrate the complementarity of the two indexes. Not surprisingly, District 12, the subject of *Shaw v. Reno,* registers the lowest score on both scales: 0.05 for dispersion and 0.01 for perimeter. By contrast, the state's other black-majority constituency, District 1, has the second lowest perimeter score (0.03) but a markedly higher, almost midrange dispersion score (0.25): not a surprising result in view of the district's comparatively compact overall shape, thicker appendages, and bulky northern core.

The complementarity of the two indexes is also evident in a comparison of Districts 4 and 6. District 6, which registers the highest dispersion score (0.44) and a low perimeter score (0.09), is slightly less elongated but notably more indented than District 4, which has the second highest dispersion score (0.40) and the highest perimeter score (0.32). District 6's less efficient boundary reflects penetration by District 12, which divides District 6 into two barely contiguous sections as it cuts through from northeast to southwest. (Fig. 3.6 provides a clearer picture of this incision.)

These indexes, which Pildes and Niemi computed for the Ninety-eighth and 103rd Congresses, confirm a broad increase in the complexity of congressional boundaries from the 1980s to the 1990s.[4] In North Carolina, not surprisingly, the mean dispersion score dropped from 0.36 to 0.28, while the mean perimeter index fell even more precipitously, from 0.30 to 0.09. Other southern states subject to preclearance also registered declines, especially in their mean perimeter scores. Even so, diminished compactness was not universal: among the forty-three states with more than one district, only twenty-one dispersion means and thirty-two perimeter means were lower in the 1990s. And the national means declined only marginally, from 0.37 to 0.36 for the dispersion score and from 0.28 to 0.24 for the perimeter score.

Curious about the locations and ethnic makeup of the 103rd Congress's least compact districts, the authors compiled a list of twenty-eight districts with low compactness scores. Because dispersion scores tended to be higher than perimeter scores across the country as well as for North Carolina, they adopted different thresholds: 0.15 and 0.05, respectively. Published in December 1993, six months after the Supreme Court's historic decision in *Shaw I*, their list proved at least moderately prescient: in the next few years the federal judiciary would strike down five of the seven black-majority districts and three of the four Hispanic-majority districts on their list. Spared were Illinois's Hispanic Fourth District, crafted by savvy federal judges whose strictly tailored boundaries addressed a compelling state interest, and two black districts near Miami that compared favorably with northeast Florida's horseshoe-shaped Third District, which eked out a 55 percent African American majority by stretching though fourteen counties.[5] Prominent omissions include Georgia's Eleventh District, struck down in 1995 in *Miller v. Johnson*, and Virginia's Third District, voided the following year.[6]

There's a pattern here, but it's not glaring. Although the two compactness measures are moderately correlated with each other, very low perimeter scores seem more likely than low-end dispersion scores to reflect unconstitutional irregularity. This is clear for two reasons: the courts struck down all five districts with perimeter scores of 0.01, but as their silhouettes in figure 5.4 illustrate, a highly inefficient boundary can enclose comparatively circular constituencies like Texas's

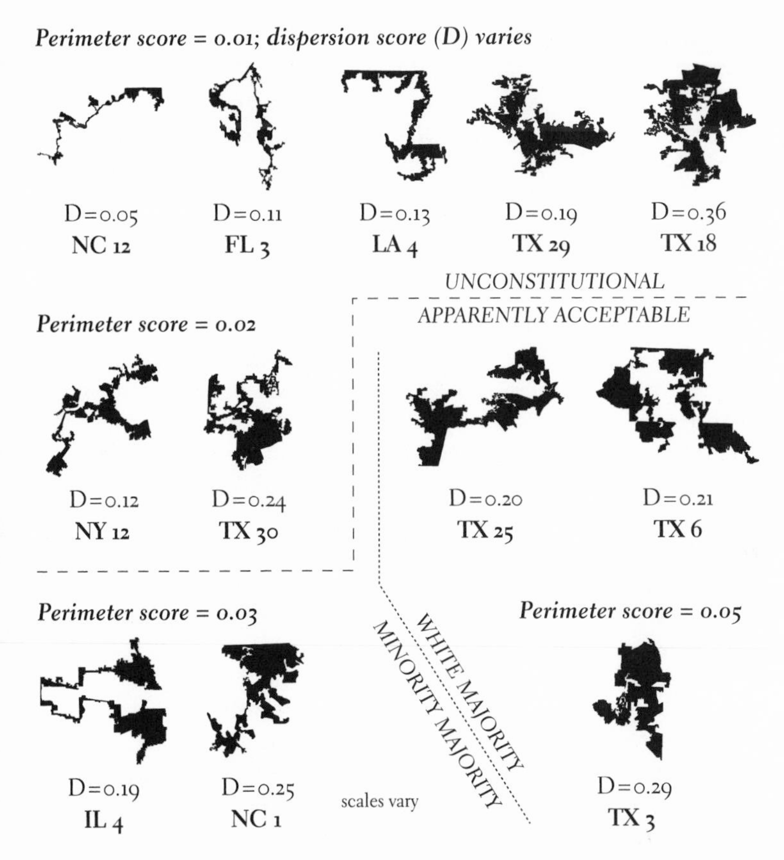

Figure 5.4. Comparatively irregular congressional districts of the 103rd Congress, ordered by compactness scores reported in Richard H. Pildes and Richard G. Niemi, "Expressive Harms, 'Bizarre Districts,' and Voting Rights: Evaluating Election-District Appearances after *Shaw v. Reno*," *Michigan Law Review* 92 (1993): 565, table 3.

Eighteenth District (dispersion = 0.36) as well as more flagrant elongations like North Carolina's Twelfth District (dispersion = 0.05). That all five are minority-majority districts reflects the difficulty of appeasing the Justice Department in regions with dispersed minority populations. By contrast, two of the four districts with perimeter scores of 0.02 had white majorities and survived whatever judicial scrutiny their inefficient boundaries might have attracted, whereas the two minority-majority districts with similarly noncompact perimeters were struck down: convincing evidence that where race has not been the prime consideration in drawing boundaries, bizarre shapes can be perfectly legal. And though neither of the two minority-majority districts with a perimeter score of 0.03 was ruled unconstitutional, both constituencies attracted convincing challenges. By inviting "strict scrutiny," an inefficient boundary makes a racial gerrymander vulnerable to judicial intervention.

Pildes and Niemi conclude that compactness measures afford an objective assessment of how far a boundary stretches the intent of the Voting Rights Act—a quantitative method akin to the disparity index for describing deviations from the one person, one vote standard. Objective assessment is important, they contend, because the "expressive harms" of bizarre shapes can "compromise the legitimacy of political institutions" by showing "disrespect for significant public values."[7] By making dispersion and perimeter scores a conscious part of the redistricting process, political cartographers might deflect cynicism if not censure by evaluating compactness numerically in the same way they consciously avoid population inequality.[8] Uniform measures are especially useful in court, they argue, because redistricting cases are rare and judges typically have little experience with the peculiarities of geographic shape.[9] Eventually, legal rules grounded in case law and judicial precedent might constructively inform negotiations between state officials and the Justice Department's preclearance mavens.

It didn't happen in the 1990s, largely because egregious, self-evident contortions like North Carolina's I-85 District and Florida's gnawed wishbone district didn't need it. Although attorneys and jurists mentioned compactness scores that supported their position, they much preferred the clever metaphors pundits eagerly concoct for highly irregular districts. In *Bush v. Vera,* for instance, Justice Sandra Day O'Connor observed that Pildes and Niemi had "ranked [Texas] Districts 18, 29, and 30 among the 28 least regular congressional districts nationwide."[10] But she quickly moved past arcane numbers to an intriguing if not inflammatory quotation from Michael Barone and Grant Ujifusa, compilers of the *Almanac of American Politics,* who compared Hispanic-majority District 29 to "a sacred Mayan bird, with its body running eastward along the Ship Channel from downtown Houston until the tail terminates in Baytown. Spindly legs reach south to Hobby Airport, while the plumed head rises northward almost to the Intercontinental. In the western extremity of the district, an open beak appears to be searching for worms in Spring Branch. Here and there ruffled feathers jut out at odd angles."[11]

In his vigorous dissent, Justice John Paul Stevens also referred to Pildes and Niemi, who had ranked six Texas districts—three white-majority districts as well as the three minority-majority districts struck down by his colleagues—"as among the oddest in the Nation."[12] And like Justice O'Connor, he preferred verbal rhetoric to numerical innuendo. Most telling is his bitterly sarcastic comparison of black-majority District 30 with white-majority District 6: a brutally effective attack (quoted earlier, on page 60) that fails to mention the areas' identical 0.02 perimeter scores.

And it's not going to happen quickly, because compactness measures as now constituted say little about how effectively a representative can represent a district. Consider, for example, North Carolina's Twelfth District, arguably the least geometrically compact district in the 103rd Congress. Bound together by I-85, this constituency is more functionally compact than its silhouette suggests.[13] By providing rapid access between the district's extremities, the limited-access highway makes the its length much less troublesome than the 160 miles reported in the media. And because the Interstate is a transport backbone, a critic's glib promise to hold rallies at exits was hardly unrealistic.[14] Moreover, District 12 is already shorter than several of the state's other districts, and I-85 not only renders it functionally compact in time and space but makes issues like transportation development, tourism, and the highway trust fund about as salient as rent subsidies and affirmative action. With these advantages, District 12 could have developed cohesiveness and identity. Far more vexing is North Carolina's white-majority District 6, cut in two by District 12 and thus only marginally contiguous. Even so, both districts, however fragile their geometry, can be defended as discernible communities of interest. And because of modern telecommunications, mass media, and the Internet, shape indexes say very little about how effectively candidates can campaign and serve constituents.

And it's not going to happen at all without complex refinements that undermine current measures' attractive simplicity. Consider the plight of Louisiana's political cartographers, forced to cope with the tortuous west bank of the Mississippi River as well as the river's classic "bird's-foot delta," through which much of the nation's runoff enters the Gulf of Mexico.[15] Redistricting committees have no control over a contorted state border guaranteed to lower the perimeter score of any congressional district its touches.[16] By contrast, a state border that follows a parallel or meridian can enhance a district's perimeter score. Louisiana, which enjoys a straight-line border with Arkansas as

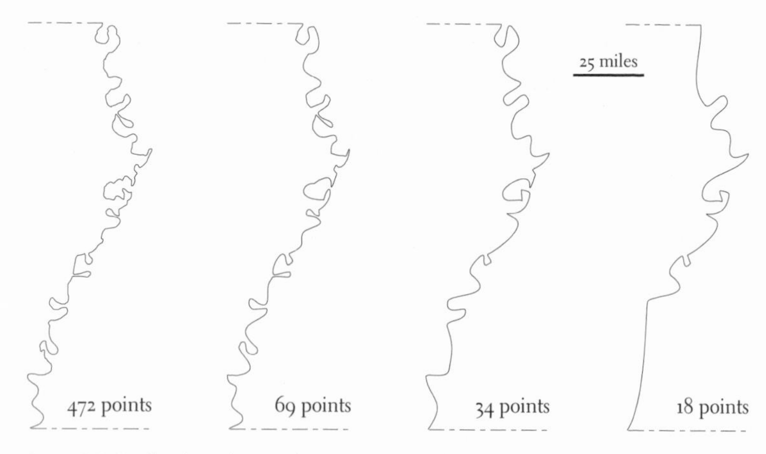

Figure 5.5. Small-scale renderings of Louisiana's riverine boundary with Mississippi demonstrate the varying degrees to which generalization algorithms can simplify, smooth, and shorten cartographic lines.

well as straight sections along its boundaries with Mississippi and Texas, squandered this geometric efficiency on the infamous "mark of Zorro" district (fig. 4.3, upper left), struck down in late 1993. The district's exaggerated perimeter includes the state's circuitous riverine border with Mississippi and numerous appendages reaching inward toward pockets of black voters. The latter irregularities are artificial and avoidable, whereas the river is not only a natural transportation barrier bridged infrequently and at considerable cost but a permanent political structure, anchored by nearly two hundred years of history. A meaningful compactness index must differentiate the inherent sinuosity of a meandering river, a contorted coastline, or a winding ridge crest from the contrived inefficiency of a line drawn to lasso blacks, Hispanics, whites, Democrats, or Republicans.

I can think of two ways to proceed: estimate the length of smooth curves that reflect the general trend of tortuous natural boundaries, or discount the measured length of these boundaries by a half, a third, four-fifths, or whatever. The former strategy is far from foolproof: cartographers have devised numerous recipes for ironing out kinks in sinuous lines, but the results can vary widely depending on the generalization algorithm used and the options or thresholds chosen.[17] Figure 5.5 illustrates a handful of the many possible simplified and smoothed renderings of the Louisiana-Mississippi border between the thirty-first and thirty-third parallels.[18] Although computerized shortening affords a more realistic estimate of a circuitous boundary's functional length, redistricting regulations would need to specify both the algorithm and its tolerances.

As for discounting the full length of sinuous boundaries, a few cartographers and lawmakers might welcome the challenge of creating an arguably objective weighting scheme that considers functional permeability, historical and political significance, and the extent to which redistricting officials can substitute other, less irregular lines. The process might employ two lengths for each boundary segment: its actual measured length and the length of a straight line connecting its endpoints. (Peninsulas, bird's-foot deltas, and other long boundaries that circle back on themselves would require two or more straight-line sections, joined at intermediate points.) A boundary following a meandering river or stream might have its length recomputed by averaging the actual and straight-line lengths, with an additional reduction applied if the feature serves as a county boundary. By contrast, actual length could be weighted more heavily for boundaries with lesser functional significance, for example, a shallow stream crossed by frequent bridges or culverts, or a ridge penetrated by tunnels. Complications abound, of course, especially if the scheme invokes discharge rates, flood probabilities, and elevations in a complex classification of drainage courses, coastlines, and ridges. More readily devised than defended, objective rules based on arbitrary weights could create more problems than they solve.

A different concern, identified by several scholars who favor numerical analysis of likely gerrymanders, is population distribution.[19] It makes little sense, they point out, to treat congressional districts as homogeneous silhouettes with uniformly spaced households. People cluster in cities and towns, after all, and a perfectly circular district with residents concentrated near its circumference is less functionally compact than an identical district with constituents clustered near the center. Even more misleading is the circular district with 90 percent of residents concentrated in a single city near its border. With much the same impact of a perimeter score that exaggerates the significance of sinuous natural boundaries, the dispersion score fails to consider an uneven distribution of population within and around the district.

Measurement advocates offer two remedies. One compares the

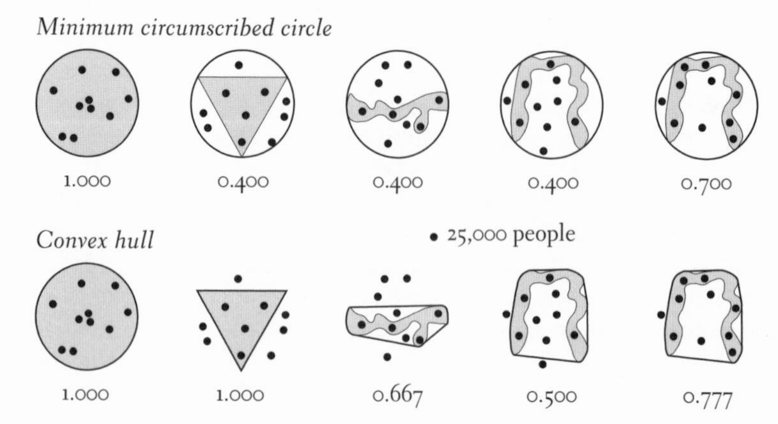

Figure 5.6. Population scores for a variety of shapes illustrate the measurement of population dispersion relative to two reference figures, the minimum circumscribed circle and the convex hull.

number of district residents with the population of a geometrically compact figure that surrounds the district, whereas the second focuses on population variations within the district.

Their first solution bases its comparisons on one of two reference figures: the *minimum circumscribed circle* (the reference figure used for the dispersion score) and the *convex hull* formed by stretching a giant rubber band around the district. Dividing the district population by the population residing within the reference figure yields a population score ranging from one for perfect population compactness to nearly zero for a district that includes relatively few residents of the surrounding region. As figure 5.6 illustrates, a completely circular district has a population score of one, whereas a horseshoe-shaped district could yield a population score close to either one or zero, depending on the number of people living in the middle.

Whether the comparison is based on a minimum circumscribed circle or a convex hull is important, especially for a district boundary tightly fitted to a naturally compact population—for instance, a long, linear valley between two other long, linear valleys. As the middle two examples suggest, an enormous circumscribed circle intersecting neighboring valleys is arguably less relevant than the hypothetical rubber band that snares excluded residents just beyond the perimeter. In general, the circumscribed circle tends to encompass a bigger area and larger population than the convex hull and thus yields a lower score. The circle might be the epitome of geometric compactness, but the convex hull provides a more relevant assessment of population compactness. Whatever the reference figure, though, the index can be radically altered by the addition of a large, contorted, and wholly unpopulated area.[20]

The second strategy treats internal population variation by invoking a concept a few readers might recall from high-school physics: the moment of inertia. In classical mechanics, the moment of inertia describes important differences among a uniform circular disk, a flywheel, and a top. With its weight concentrated near the center, the top has the lowest moment of inertia, whereas the flywheel, with most of its mass toward the edge, has a higher moment, and the disk falls somewhere in between. This concept can be applied to population—think of people as units of mass—by first determining a congressional district's center of population and then summing up for each city, town, census tract, or voting precinct the product of population and the squared distance from the population center.[21] In general, the smaller the result, the closer the district's residents to the center of population and the greater its compactness. Dividing by the hypothetical moment of inertia computed for the same population evenly distributed throughout the district yields a ratio less than one for a relatively compact district and greater than one for a comparatively dispersed population. A score comparable to measures ranging from zero to one is also possible: merely divide the actual population moment by the hypothetical moment of a population uniformly distributed along the district's perimeter.

However meaningful to mechanical engineers, the population moment of inertia has few advocates among redistricting experts, who seem baffled by its interpretation. The index involves unfamiliar terms, contradicts public understanding of spatial compactness, and like the population-comparison score, is unpredictably distorted by unpopulated territory. In an assessment of comparative measures that merit consideration by redistricting officials, David Horn of the Center for Research into Governmental Processes and his colleagues rejected population moments as "unsound and complex."[22] And Pildes and Niemi, who considered the Supreme Court's concerns, dismiss all population measures as irrelevant to issues raised in *Shaw*.[23] If considered at all, population might be best be presented on a map.

Some political theorists question the relevance or appropriate role of formal, measurable criteria like compactness, contiguity, and population equality. Andrew Gelman and Gary King, for instance, argue that shape per se is not a reliable indicator of vote dilution, while Micah Altman contends that the Supreme Court distorted the process of redistricting by placing an "extreme emphasis" on easily measurable mathematical criteria.[24] Others reject as futile the search for functionally effective formal criteria.[25] Despite these reservations, shape indexes and other measurements are unlikely to disappear as long as judges are willing to look at maps and disparity ratios.

6 *Props and Propaganda*

MAPS ARE INDISPENSABLE IN REDISTRICTing, not just as a frame for redrawing boundaries but as a medium for informing legislators, the public, and the courts. Consumers and critics of democratic representation—you and I, party politicians and three-judge panels, leaders and members of protected groups, and anyone else who votes or cares—need to know not only what district we're in but how well the political cartographers have served our interests (and their own) while satisfying constitutional and statutory mandates. Early in the process, when officials are holding hearings or debating alternatives, maps become tools of persuasion if not weapons of ridicule. Before boundaries become official, they are on trial in the media and at public meetings, where lines and labels testify to a district's geographic integrity: how well the new partitions reflect communities of interest, keep counties intact, and let us vote again for the incumbent who once helped us fend off the federal bureaucracy. Or, as happens, how outrageously the politicians' self-serving borders meander hither and yon to corral blacks (or Republicans), marginalize white Democrats (or Hispanics), or protect the seat of that wrong-headed charlatan who's been hoodwinking voters for years.

As this chapter observes, maps are generalizations of reality and intrinsically rhetorical. Because large geographic areas are compressed into small cartographic spaces, clarity demands that most features be suppressed; otherwise the map becomes hopelessly cluttered and useless. This need to simplify authorizes the mapmaker to select what's relevant, suppress what isn't, and if he or she chooses, craft a convincing but biased view of reality.[1] Not that all map authors are deliberately devious: most cartographic bias, I'm convinced, arises from sloppy compilation or slavish deference to tradition. Silhouette maps seem quite all right (Aren't they?) because Supreme Court justices use them, not to mention professors of geography. Despite this potential for manipulation, maps enjoy enormous credibility with a public not accustomed to questioning their crisp, professional-looking lines and labels. Location is fact, maps show location, so maps are factual. Perhaps, but maps like words can spin the interpretation of some facts while diverting our attention from others.

The map's power to distract is particularly apparent at public hearings like one I attended in 1992.[2] New York State's Legislative Task Force on Demographic Research and Apportionment had just fashioned new boundaries for the state's assembly and senatorial districts, and a contingent of appointed officials and staff members had hit the road for hearings in several upstate cities, including Syracuse. Scheduled for late Saturday morning at the Onondaga County Court House in Syracuse, the hearing was to last until all citizens had been heard. Attendees could pick up a free book of maps, sign the register, and critique the new lines. Task Force members, who had commandeered the legislative chamber for the day, listened patiently as a parade of twenty-eight citizens, mostly local politicians, stepped to the microphone and denounced boundaries that would split their counties, disrupt the constituency of a beloved incumbent, or lump together two small cities that had thrived for decades as centers of separate assembly districts. Most speakers pointed fleetingly to the regional boundary maps displayed before the dais on two huge easels: key props in a

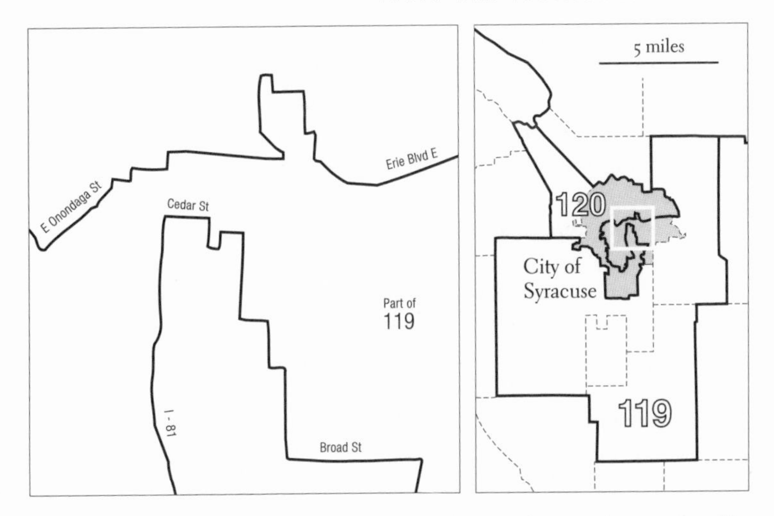

Figure 6.1. Excerpt (left) from map showing the division of the city of Syracuse between Assembly Districts 119 and 120, which interlock in a coupler-like grip (right) that concentrates minority voters in District 119. Left-hand map from New York State Legislative Task Force on Demographic Research and Reapportionment, *Co-chairmen's Proposed 1992 Assembly and State Senate District Boundaries, January 21, 1992,* 40.

pseudopopulist ceremony that ignored the need for election reform, responsible budgeting, and wider use of public referenda.[3]

Barely intelligible beyond the first few rows, the maps on display were no less useful at the back of the room than the minimalist urban maps in the Task Force atlas. On the left side of figure 6.1, an excerpt from the map of assembly districts proposed for the city of Syracuse shows jagged district boundaries that generally follow local streets, few of which are labeled or readily recognizable. The only landmarks are the municipal boundary and portions of arterials like Interstate 81, which contributes fleetingly to the border between the 119th and 120th Assembly Districts. That the map could be enlarged and superimposed onto a conventional street map was little consolation to Syracuse residents attending the hearing, few of whom spoke up. Ironically, most speakers assailed the Task Force's work in nearby less populated counties, where districts divided along town and county boundaries yielded relatively intelligible maps.

Equally inaccessible were the "finely detailed, high-resolution 3 × 4 foot mapsheets [drawn] to the street level" that the Task Force deposited in seventeen libraries throughout the state.[4] Larger than the typical atlas or topographic map, the sheets were not easily filed. Several months later I checked the central libraries in Syracuse and in Utica, a small city fifty miles east, and found that few citizens had bothered to consult the oversize maps, which librarians had rolled together and placed atop a row of filing cabinets in the reference department. Each time someone unfurled the bundle in search of a specific chart, the cumbersome heap become more disorganized. And once elections were held, library officials gladly tossed out the whole mess. Local newspapers further testified to the maps' inadequacy by plotting district boundaries on their own more generously informative local base maps.

More revealing were the numbers at the back of the book: population counts for individual districts as well as totals and percentages for six racial or ethnic groups. According to these tabulations, the proposed 119th Assembly District, where I lived, contained 115,475 people, 4,461 (or 3.72 percent) fewer than the statewide average. By race and ethnicity our district was 68.81 percent non-Hispanic white, 24.72 percent non-Hispanic black, 2.91 percent Hispanic, 2.37 percent Asian, 0.96 percent Native American, and 0.23 percent "other," making us a minority-influence district of sorts and a nice mix for the Democrat incumbent. By contrast, the Task Force offered the Republican incumbent in the neighboring 120th Assembly District an equally safe 91.17 percent white majority, largely by winding the two districts together in a couplerlike grip (fig. 6.1, right) familiar to anyone who ever worked on a railroad or fancied model trains. (If you cook or enjoy sweets, the map might also suggest a swirl cookie.)

Despite this suggestion of a benign interest in the rights of minority voters, New York's openly bipartisan political cartographers are committed primarily to protecting incumbents. And they've become good at drawing boundaries that help the Democrats control the assembly while Republicans dominate the state senate. It seems ironic, then, that the clearest, cleanest map in the Task Force atlas (fig. 6.2) beckons the public to "Help Us Draw the Lines!" by attending one of the twelve public hearings.[5] As if the promise of "public participation" guaranteed nonpoliticians a significant role in crafting more compact, less partisan districts.

The next edition of the Task Force atlas, this time showing districts as "approved by the Legislature," included numerous changes, mostly small ones, some possibly reflecting comments taken at the hearings.[6] I noticed a few minor adjustments to boundaries within

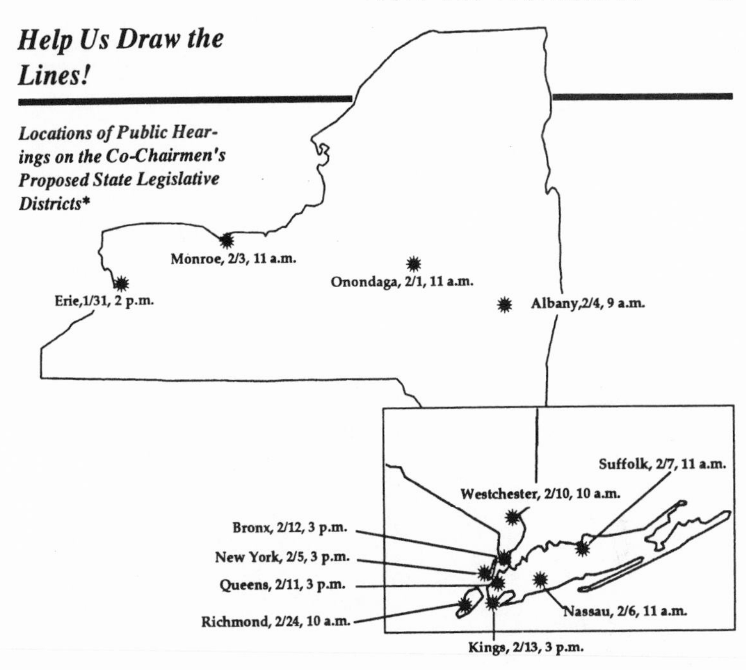

Figure 6.2. An invitation to wannabe political cartographers. From New York State Legislative Task Force on Demographic Research and Reapportionment, *Co-chairmen's Proposed 1992 Assembly and State Senate District Boundaries, January 21, 1992*, 9.

Onondaga County, where a net gain of four people helped Assembly District 119 advance from 24.72 to 25.01 percent black while a net loss of four people dropped District 120's black percentage from 4.55 to 4.26. Someone apparently identified additional black neighborhoods and some white ones as well, for the change reflected a net transfer between districts of 336 African Americans. This exchange brought the state assembly plan into slightly greater accord with the Task Force's objective of "strict adherence" to the Voting Rights Act.[7] As the group's cochairmen noted in the atlas's preface, "Even when it is not possible to create minority seats, minority communities throughout the state were 'kept whole'" by consolidating smaller minority communities like that in Syracuse "into districts which maximize minority influence."[8] If I didn't know better, I might think race was a greater concern than incumbency and partisan interests.

Next time around, I'm certain, the Task Force will provide far bet-

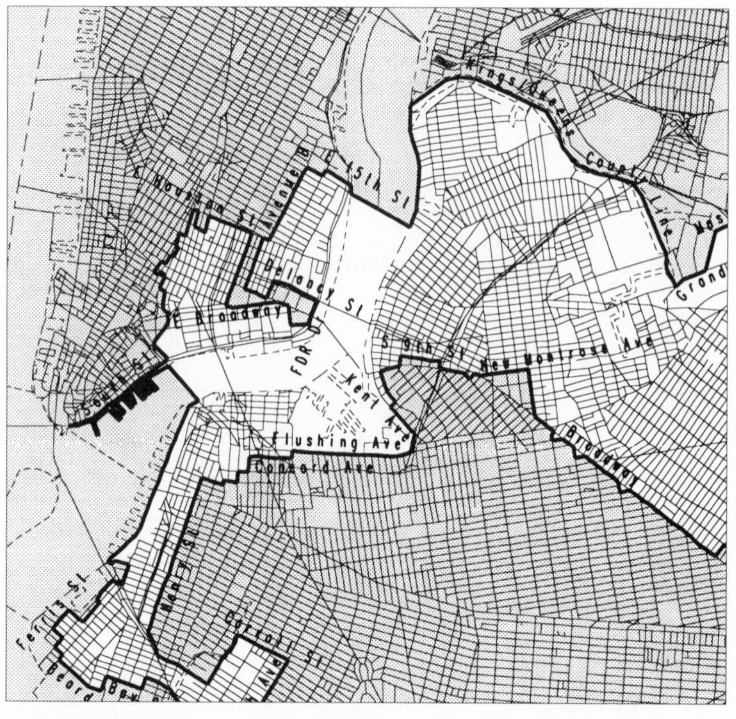

Figure 6.3. Excerpt from New York State Legislative Task Force on Demographic Research and Reapportionment map of Congressional District 12, ca. 1997.

ter maps. Its cartographic technology is better now, as the agency demonstrated in describing the redrawn boundaries of the Bullwinkle District and its neighbors. Figure 6.3, an excerpt from a Task Force map of Congressional District 12's new borders, redrawn in 1997, shows every street and block as well as assorted piers along the East River and substantial streetless areas that are largely nonresidential. A detailed street grid and labels for most bounding streets help residents who know the area find their blocks. By revealing the political cartographer's building blocks, the new maps afford a more intelligent assessment of geography and shape than either the cryptic sketches handed out at Task Force hearings in 1992 or the silhouette maps appended to recent Supreme Court opinions.

What's so wrong with silhouette maps? Lots of things. As I noted in the preceding chapter, these stark black-and-white drawings

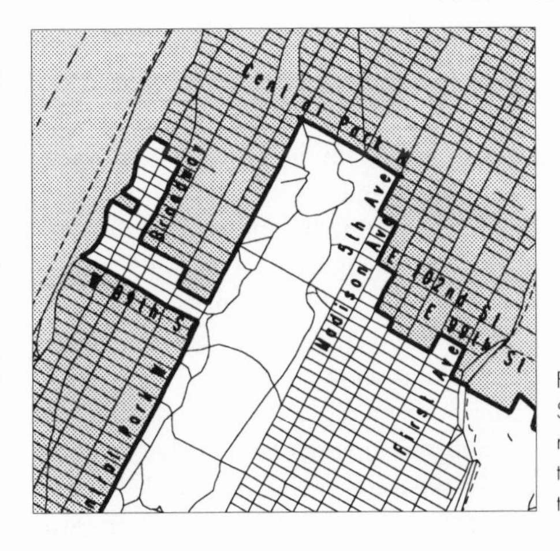

Figure 6.4. Excerpt from New York State Legislative Task Force on Demographic Research and Reapportionment map of Congressional District 14, ca. 1997.

not only ignore internal variations in population density but highlight sinuous borders that might be perfectly logical or unavoidably contorted. In suppressing shades of gray, silhouettes imply a misleading uniformity and focus undue attention on compactness. To properly assess the relevance of boundaries in urban areas requires detailed, block-level maps showing population density, minority composition, and land use. Even a detailed street map can show how political cartographers use uninhabited land to stretch the notion of contiguity. Look, for instance, at figure 6.4, in which New York City's Fourteenth Congressional District reaches more than half a mile across Central Park to link affluent East Side neighborhoods with two dozen well-to-do Upper West Side blocks. A strict sense of contiguity would favor adding blocks east of the park along the present boundary—an alternative conveniently suppressed by a silhouette that masks uninhabited terrain and stops at the district line. In this case the cartographic silences just beyond the boundary would be as misleading and onerous as those within.[9]

Land uses that exclude or limit human habitation are also relevant in rural areas, where vast parcels of federal- or state-owned park, forest, and range lands can link otherwise noncontiguous settlements while making a district border appear highly irregular or comparatively smooth. To show this properly almost always requires more space than is allotted silhouette maps in newspapers, extremist tracts, and recent Supreme Court opinions. It's folly to expect a meaningful assessment of alternative configurations from a map two or three inches across.

Equally troublesome are large bodies of water, omission of which can make legislative districts in lake-rich states like Florida, Minnesota, and Wisconsin resemble Swiss cheese. It seems reasonable for silhouettes to swallow inland lakes, especially in Texas, where a generous interpretation of the contiguity requirement allowed real, fully inhabited cavities that resemble hydrographic inclusions. (My reduced, more generalized renderings in figures 4.4 and 4.5 omit a number of the smaller enclaves readily apparent in the silhouette maps appended to opinions in *Bush v. Vera.*) More problematic are coastal areas, where bays, sounds, and wide rivers separate parts of districts linked by a bridge, a ferry, or merely a line of sight. Should the boundary follow the riverbank and cross along a bridge, suitably widened for the occasion, as in my portraits of the Bullwinkle District (figs. 1.2 and 4.6), or should the mapmaker assign all parts of an estuary to some district, thereby mollifying jagged boundaries with straight lines and inking in indentations in their silhouettes?

If you think there's a right answer, you'll be disappointed. In the Census Bureau's *Congressional District Atlas,* lines cross rivers and lakes, but only to link boundaries on opposite shores; the *Atlas* scrupulously avoids silhouettes. At a considerably smaller scale in *The Historical Atlas of United States Congressional Districts, 1789-1983,* political geographer Kenneth Martis described congressional districts with similarly simplified boundaries that enclose and otherwise ignore all but the widest estuaries and lakes.[10] By contrast, the *New York Times* distinguishes land from water on its maps of New York's Twelfth Congressional District,[11] and my Bullwinkle-like silhouettes in figures 1.2 and 4.6 follow suit by omitting the East River—hardly a coincidence considering this book's title. Although readers might at least expect consistency within an atlas, a book, or a newspaper, whether to omit estuaries or fill them in is largely a matter of cartographic license, which understanding viewers, aware that mapmaking is as much an art as a science, willingly tolerate.

Silhouettes and other highly simplified maps seem to trigger an extreme form of map generalization: the substitution of words for images. Chosen to ridicule the district or impugn its perpetrator, the words are not only graphic but pejorative. The prototype is, of course, *gerrymander,* which needs no accompanying artwork to evoke the mental image of a writhing reptile.[12] I've searched the cartographic literature for discussion of verbal substitution but found nothing—

generalization scholars have yet to recognize silhouette maps as geographic Rorschach tests. But the phenomenon is very real to Deval Patrick, assistant attorney general for civil rights under Bill Clinton and Janet Reno. Writing in the *Emory Law Journal,* Patrick critiqued the cartographic weapons that punctured the government's case in *Shaw v. Reno:*

> Let us look at the charge that these districts are "bizarre." We have all seen the pictures, and we have heard the epithets: "snakelike," "earmuffs," "spiderlike," and "the mark of Zorro." I will admit that some of these districts look pretty strange to me as well, until you look around. A lot of majority-white district are bizarre and these districts have not been affected by the Voting Rights Act.[13]

The weirdness, he maintained, reflects less the rush to pack African Americans or Hispanics into minority-majority districts than the legislature's eagerness to protect its own. "There's no such thing as a 'normal' or regularly shaped district," he argued.[14]

To demonstrate that cartosilhouettes are a double-edged sword, Patrick cited a few districts and appended their maps.

> If you flip through a reference book like the *Almanac of American Politics* and take a close look at congressional district maps, it is practically impossible to find one that is of a regular shape. I include some of my personal favorites. The one from Texas [Congressional District 6] is about as "bizarre" as any district you can imagine. Consider the district in Massachusetts [District 4], and how it bears a strange resemblance to a saxophone. The one from North Carolina [District 10] is rather odd as well....These districts have populations that are, on average, ninety percent white, yet they are among the ugliest, the most bizarre, the least compact in the country, but nobody ever accuses them of violating the Constitution.[15]

Yet in the hands of conservative white lawyers, silhouettes apparently cut in only one direction.

That's clearly the thrust of Robinson Everett, plaintiffs' attorney in *Shaw v. Reno.* Commenting to a *New York Times* reporter on Justice Sandra Day O'Connor's majority opinion, Everett maintained that "strange shapes or bizarre shapes are significant because they show why the district was created; they illustrate its purpose...[a] quota system in the election of members of Congress."[16] This notion that form follows function led Justice O'Connor, in her majority opinion, to denounce the I-85 District's shape and question its constitutionality. Simplified maps proved equally persuasive three years later, in *Bush v. Vera,* when she presented the Court's decision to strike down a Texas district because of a "combination of a bizarre, noncompact shape and overwhelming evidence that the shape was essentially dictated by racial considerations of one form or another."[17] To reinforce her point, O'Connor appended silhouettes of the three highly irregular and now unconstitutional minority-majority districts, while Justice John Paul Stevens countered in his dissent with silhouettes of three similarly contorted white-majority districts.

Some observers dispute Justice O'Connor's finding that race was the predominant cause of the irregular boundaries. In an eloquent essay titled "Gerrymanders: The Good, the Bad, and the Ugly," former Stanford University law school dean John Hart Ely blamed Democrats in the Texas legislature, not the Voting Rights Act.

> One thing on which there seems to be consensus, however, is that the most bizarre district shapes are seldom caused by a desire to create majority-minority districts [which] can usually be accomplished without inordinate frenzy. Rather, the zaniness results from a tortured interaction of ethnic and more directly political concerns—that is, from the process of creating a district calculated both to elect a minority and at the same time to control the damage to the Democrats (and, what may be more important even to Democratic legislators and certainly is to their Republican colleagues, to make sure that no incumbent's seat is put at risk).[18]

In situations like these, he argued, no one can disentangle race and incumbency, and the "dominant purpose" test outlined in *Miller v. Johnson* is not only "vague and manipulable" but "ultimately incoherent."[19] Suspicious of claims that silhouette maps reveal motive, Ely carried O'Connor's distaste for weird shapes one step further in concluding that bizarreness by itself seems the best criterion for striking down outrageous gerrymanders of any type, racial or political.[20]

Maps have no business in high court opinions, insists legal scholar Hampton Dellinger, who would severely limit their use. In a *Harvard Law Review* article titled "Words Are Enough: The Troublesome Use of Photographs, Maps, and Other Images in Supreme Court Opinions," Dellinger attacked "the Court's undistinguished use of visual attachments and their inherent susceptibility to manipulation."[21] His cartographic examples reach back to 1946, when the plaintiff in *Colgrove v. Green* challenged the severe numerical disparity among Illinois congressional districts. In presenting the Court's decision not to intervene, Justice Felix Frankfurter attached maps of Illinois and three other states with thickets of obviously gerrymandered districts.[22] According to Ely, Frankfurter's maps not only drew attention away from the outrageous numerical disparities—the Illinois districts ranged in population from 112,000 to 914,000—but explained the justices' reluctance to attack a widespread problem with no ready solution.

Fourteen years later Justice Frankfurter injected cartographic propaganda into another landmark decision, *Gomillion v. Lightfoot.* White officials in Tuskegee, Alabama, had redrawn the municipal boundary to exclude all but four or five of the city's 400 black residents, who lost services as well as their right to vote. In Dellinger's view, this was a clear case of illegal racial discrimination, and Frankfurter had no need to attach a before-and-after map (fig. 6.5) dramatizing Tuskegee's radical transformation from a square to a twenty-eight-sided polygon that excluded not a single white voter. Constitutional issues aside, the Court's crude map explains why the justices could not ignore this outrageous redistricting.

Two decades later, in *Karcher v. Daggett,* the high court began to read more meaning into maps. The case involved population equality, and Justice William Brennan, who wrote the majority opinion, said nothing about the attached district map.[23] But in a separate, concurring opinion Justice John Paul Stevens observed that the "shape of the district configurations themselves" was no less important than the statistical evidence.[24] And his dissenting colleague Justice Lewis Powell observed that "the map attached to the court's opinion illustrates [unconstitutional gerrymandering] far better than words can describe."[25] Three years later, in *Davis v. Bandemer,* Justice Powell inserted two maps, compared the shape of one district to a salamander, and suggested that "in some cases proof of grotesque district shapes may, without more, provide convincing proof of unconstitutional gerrymandering."[26] To reinforce the point, he cited the attachments in *Gomillion* and *Karcher.*

These sentiments proved decisive in *Shaw I,* when Justice O'Connor attached a map of North Carolina's twelve districts, repeatedly attacked "boundary lines of dramatically irregular shape,"[27] and argued vehemently that "reapportionment is one area in which ap-

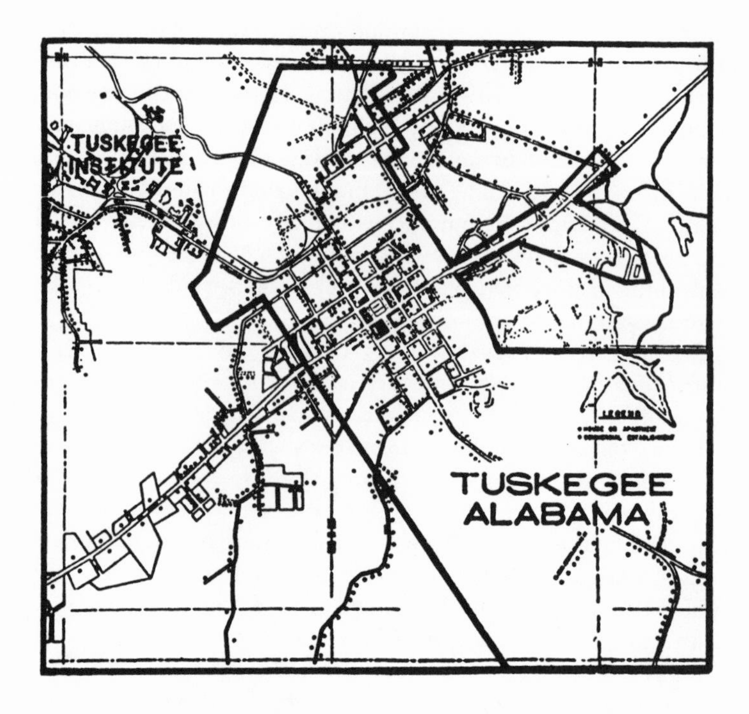

Figure 6.5. Map of Tuskegee, Alabama, appended (at this size and no more legibly) to the Supreme Court's decision in *Gomillion v. Lightfoot*. According to the caption, "The entire area of the square comprised the City prior to Act 140. The irregular black-bordered figure within the square represents the post-enactment city." From *Gomillion v. Lightfoot*, 364 U.S. 339, 348 (1960).

pearances do matter."[28] Three years later, in *Shaw II*, Chief Justice William Rehnquist pointed to the earlier attachment and asserted that "a map portrays the districts' deviance far better than words."[29] The Court deemphasized shape in *Miller v. Johnson* because the Georgia district in question was not highly irregular, but compactness reemerged as a key issue in *Bush v. Vera* when Justice O'Connor inserted silhouettes of three districts and attacked their shape. By making compactness the focal point of the Court's decision, her maps weakened the majority position that the true evil was a predominant racial motive.

Dellinger's prescription is a ban on embellishing high court decisions with vague, very small scale, "decontextualized" boundary maps that say little about how representatives might interact with constituents.[30] Far more relevant, he argues, are city street maps and state road maps, which reflect natural barriers and social connections. What's more, *United States Reports* and other publications of Supreme Court decisions should identify the party that introduced a map as evidence, indicate the original size of any illustration reduced to fit the page, and never reproduce color maps in stark, pejorative black-and-white. Aware that the justices and their publishers are unlikely to reform the process, he counsels informed skepticism: "When confronted with an attachment, readers should ask: Why has a Justice included it? Is it an atypical perspective? Is critical objective information omitted, and if so, why?"[31]

7 *Immunizing Incumbents*

SAM STRATTON'S DEAD NOW, BUT HIS SKILL in winning elections as a Democrat in heavily Republican districts is legend in upstate New York. I didn't like his politics—Stratton was an ardent supporter of the Vietnam War and military spending—but I had to admire his popularity and persistence. For this chapter, focused on incumbency, his fifteen-term career in Congress provides a dramatic illustration of how political parties use redistricting to protect members and eliminate opponents.[1]

A naval intelligence officer awarded the Bronze Star for combat service in World War II, Stratton entered politics in 1950 in Schenectady, New York, where he served two terms on the city council.[2] Elected mayor in 1955 by a meager 282-

vote margin, he fulfilled his promise to clean up city hall by leading a police raid on a gambling den several doors away. Name recognition was never a problem: because Schenectady didn't pay its mayor a full-time salary, Stratton kept his job in local radio and television as an announcer, newscaster, and kiddie-show performer. Years later many voters remembered him as the straight-shooting, harmonica-playing cowboy Sagebrush Sam. Few viewers knew of his master's degrees in philosophy from Haverford and Harvard.

Stratton ran for Congress in 1958 and impressed the *New York Times* with a 13,000-vote plurality: a remarkable victory for an upstate Democrat and an embarrassing defeat for Republicans, who had held the seat for forty-two years.[3] Although Schenectady County gave him a 10,000-vote lead, Stratton ran well throughout the Thirty-second District, comprising counties within the Schenectady media market to the west and north (fig. 7.1, upper left). He was also lucky: the district's longtime Republican incumbent had decided to retire, and Stratton's lesser-known opponent was merely secretary of the Assembly Ways and Means Committee.

State Republicans, who controlled the assembly, the senate, and decennial reapportionment, hoped to regain the seat after the post-1960 remap. What better way to unseat Stratton than to combine Schenectady and Albany Counties into a single district and force him into a primary with fellow Democrat Leo O'Brien, who had represented Albany County in the House since 1952? In case Stratton decided to join the remainder of his loyal constituents, Republican mapmakers configured the adjoining Thirty-fifth District (fig. 7.1, middle left) to ramble westward from Amsterdam 170 miles through the southern reaches of the Utica, Syracuse, and Rochester media markets, where few voters had ever heard of Sagebrush Sam.

Forced to choose between early retirement, political suicide, and moving to Amsterdam, Stratton filed for reelection in the Thirty-fifth District, intended as a safe seat for Republicans, who outnumbered Democrats two to one in voter registrations.[4] As before, Stratton was lucky: the conservative Republican who had represented the area for forty years had just stepped down, and his intended heir was a state senator little known outside two or three counties. And as in 1958, the tenacious Democrat's 12,000-vote margin was a major upset for the GOP. Quoted in the *New York Times,* Stratton applauded voters for repelling the Republican effort "to eliminate a legislator by drawing lines on a map."[5]

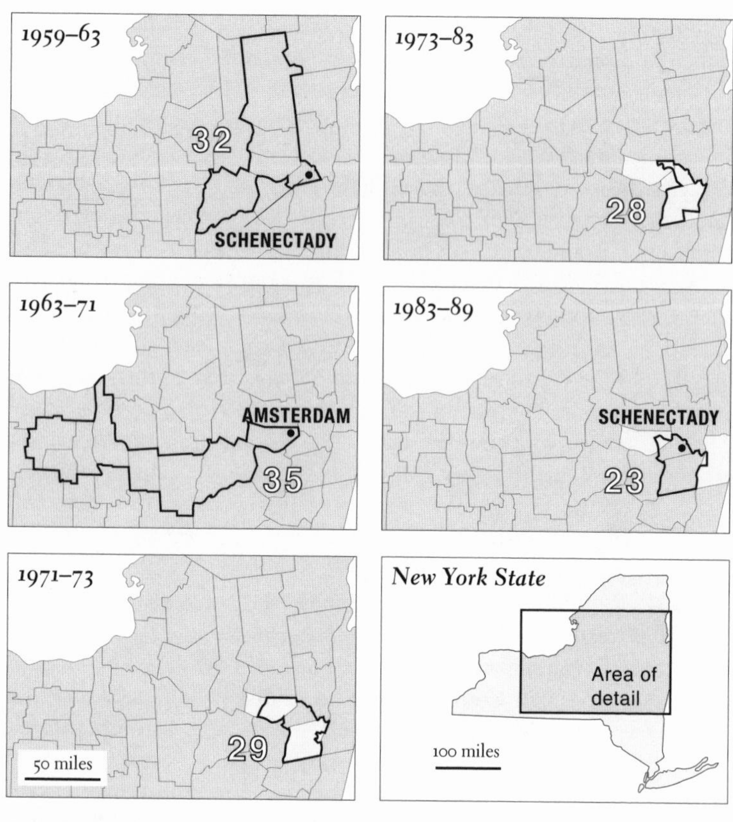

Figure 7.1. Stratton's constituencies. Between 1959 and 1989 Democrat Samuel S. Stratton represented five different congressional districts and moved his residence from Schenectady to Amsterdam and back again. Stratton's districts did not include split counties (shown in light gray) until 1971, after an interim redistricting to satisfy the one person, one vote requirement.

In the parlance of political scientist Leroy Hardy, Stratton was the intended victim of an "elimination gerrymander," designed to force incumbents from the opposing party to fight over a single seat.[6] Because one incumbent must drop out or lose, the ploy opens up a seat for the party drawing the lines. What's more, both incumbents could disappear if the loser in the primary drains votes from the winner by running anyway as a minor-party candidate or an independent. Although mapmakers occasionally don't care who loses or bows out, they often put a prime target like Stratton into a new district with voters largely loyal to another incumbent. And for good measure they divide the intended victim's old district among several constituencies.

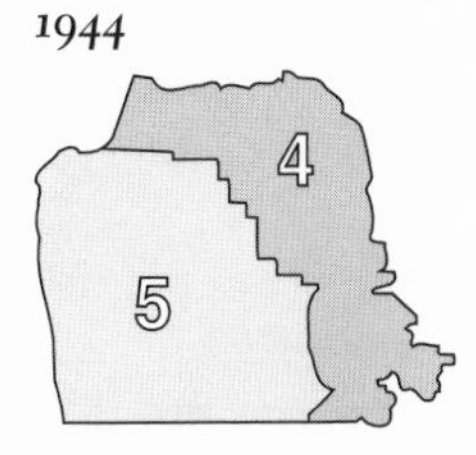

Democrat Frank Havenner wins in District 4 and holds seat for four terms.

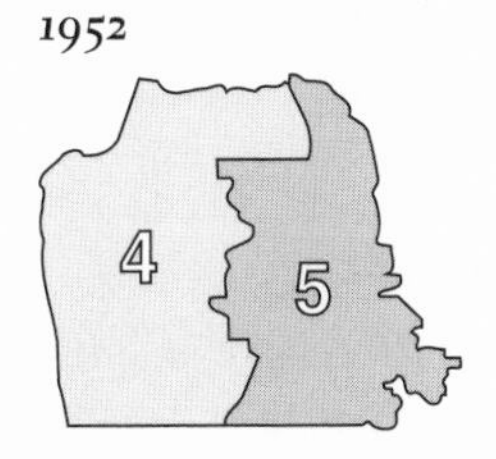

Havenner loses redrawn District 5 to Republican William Mailliard.

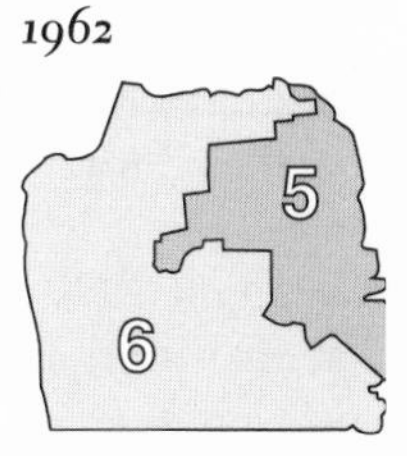

Although reconfigured Democrats, District 5 st reelects Mailliard.

Figure 7.2. Elimination gerrymandering in San Francisco. The tactic worked in 1952 (center), when Republicans undermined the reelection of a liberal Democrat. But lines drawn ten years later to favor a Democratic challenger (right) failed to unseat the Republican incumbent.

Should the mark relocate, as Stratton did, voters typically cooperate by choosing the more familiar candidate. But as Sagebrush Sam demonstrated, the elimination gerrymander can be risky when the dominant party runs a less seasoned candidate against an experienced, highly personable incumbent.

When slow growth costs the state a seat or two, the party in power might even finger one of its own. As Hardy observed, when "someone has to go, it is sometimes surprising how readily incumbents can agree on who it should be—first, anyone but me, and, second, the member who generally bothers his colleagues for one reason or another. Often the agreement is bipartisan."[7]

Elimination gerrymandering might seem unavoidable in New York and other states with declining populations, but Hardy describes several instances in California, which routinely gains seats.[8] San Francisco at midcentury provides an intriguing example (fig. 7.2). The city had two congressional districts, separated after the 1940 census by a northwest-southeast boundary. District 4, northeast of the line, included the comparatively liberal Pacific Heights and Marina neighborhoods and tended to favor Democrats. In 1944 the district elected Frank Havenner, a liberal Democrat who won the next three elections. Unfortunately for Havenner and several like-minded colleagues, the Republicans who controlled redistricting in 1951 targeted liberal Democrats throughout the state. To purge Havenner, they rotated the district boundary about forty degrees clockwise, swapping the northern part of his old district for more conservative middle-class precincts farther south. As expected, the incumbent lost to Republican challenger William Maillaird by a decisive 18,000 votes.[9] Ten years later, when Democrats controlled the legislature, their reconfiguration of San Francisco's districts failed to unseat the now-entrenched Maillaird. As Sam Stratton demonstrated, elimination gerrymanders don't always work.

In the 1990s some incumbents faced a new danger: pressure to create new minority-majority districts. This threat was especially acute in New York, which added many Hispanic residents during the 1980s but lost three seats in the 1990 enumeration.[10] Black and Hispanic activists lobbied redistricting officials and the Justice Department for maximum representation, but legislators could not agree on which incumbents to eliminate. As the deadline approached, only one House member had volunteered to retire, and potential candidates were confused about where to run and when to file. Unwilling to let the old boundaries stand, a three-judge federal panel appointed a "special master," whose plan would have forced eight senior House members into races with other incumbents. Wary of lost clout—senior representatives dominate congressional committees, and the committees dispense pork—legislative leaders agreed on a map designed to minimize incumbent-incumbent contests.[11] The plan included four black-majority districts, all with African American incumbents, and only one new Hispanic-majority district. Soon after the map was published two more representatives retired, and in a quixotic act of resistance, veteran Brooklyn Democrat Stephen Solarz, perhaps thinking his surname might save him (Sounds Latino, doesn't it?) volunteered to run in the newly created Hispanic-majority Bullwinkle District.[12] Despite $2 million in campaign funds, he finished second in the Democratic primary.

Sam Stratton became a target a second time in 1970, when the Republican legislature drastically reconfigured his district. GOP mapmakers normally would have waited for results from the 1970 census, but in April 1969 the Supreme Court had ordered New York to redraw its congressional districts for the 1970 elections.[13] Although population counts were nine years old, the high court was fed up with states reluctant to implement the one person, one vote doctrine extended to congressional districts in 1964 in *Wesberry v. Sanders.* New York was one of the footdraggers: in 1968, in response to an earlier court order, the legislature had adopted a redistricting plan with a 14 percent population disparity. (Democrats controlled the state assembly briefly in the mid-1970s, and Stratton's district remained intact.[14]) The high court declared the remap unconstitutional and referred the case to a three-judge panel, which set a not too generous deadline of January 30, 1970—sufficient time for a court-appointed special master to do the job.[15] Should they fail again, legislators risked a wholesale slaughter of incumbents, Republicans as well as Democrats.

Handed another opportunity to divide and conquer, Republican mapmakers once again separated Stratton from his constituents. The new attempt at elimination was more vicious than the first. Encouraged by recent court decisions to split counties in the interest of population equality, GOP strategists not only reassigned seven of the eight counties that Stratton had represented in District 35 (fig. 7.1, middle left) but left him only half his home county.[16] In addition to pulling in Schenectady County, which he had not represented since 1963, they added most of Albany County—but not the city of Albany's traditionally Democratic ethnic and inner city neighborhoods (fig. 7.1, lower left). And this time he faced a congressional incumbent, two-term Republican Daniel Button, a former newspaper editor. With 85 percent of his old district intact, Button had an overwhelming advantage.[17]

Once again the GOP misjudged Stratton's charisma. By all accounts the tall, trim, dynamic former telecaster was a better campaigner than his comparatively short, portly, soft-spoken opponent.[18] Perhaps the election was a referendum on social spending and national defense: the hawkish Stratton was fiscally conservative, whereas the more liberal Button actively supported federal aid for education, housing, and health care. Maybe voters resented cartographic manipulation by the state legislature or meddling by the *New York Times* editorial board, which endorsed Button.[19] For whatever reason, early returns revealed a strong lead for Stratton, and Button conceded defeat less than an hour after the polls closed.[20]

A few weeks after Stratton returned to Washington to represent the Twenty-ninth District, the Census Bureau confirmed New York's loss of two House seats. Although cities throughout the state had declined during the 1960s, severe out-migration from New York City placed downstate incumbents at greater risk than their upstate colleagues. Even so, the twice-targeted congressman from Amsterdam was wary of cartoassassination. But this time he and his Democratic colleagues had found an ally in Republican governor Nelson Rockefeller, who was eager to barter cartographic clemency for a federal bailout. New York was suffering its biggest economic setback since the Great Depression, and the governor hoped to rescue his presidential aspirations as well as the state's finances.[21] Help us, Rockefeller offered, and we'll help you. The deal worked: Congress passed revenue sharing, Republican mapmakers were uncommonly cooperative, and Congressman Stratton got the new, slightly smaller Twenty-eighth District (fig. 7.1, upper right), which included all of the cities of Albany and Schenectady. Despite Richard Nixon's 1972 reelection landslide, Stratton trounced his conservative Republican opponent four to one.[22]

For upstate congressional Democrats the post-1980 remap was even easier. In 1974 their party captured the state assembly, largely because of the Watergate scandal, which drove Nixon from office, launched the once-popular Rockefeller into the vice presidency under Gerald Ford, and soured voters on Republican candidates at all levels.[23] Congressional redistricting was a joint responsibility of the state assembly and the state senate, and Republicans lacked the power if not the incentive for elimination gerrymandering. The Watergate backlash also helped elect a Democratic governor: former congressman Hugh Carey, whom voters reelected in 1978. Able to veto redistricting bills and possibly provoke the courts into appointing a special master, the highly partisan Carey helped ensure the congressman from Amsterdam another safe seat: the Twenty-third District, configured largely from familiar territory (fig. 7.1, middle right). Secure for the next decade in his new district, Stratton, now in his middle sixties, moved back to Schenectady.[24]

Like other large states, New York faced a complex and potentially contentious post-1980 remap. Fueling the judiciary's growing intolerance of population disparities were the Census Bureau's finely detailed block-level data, which made it easy to attain nearly exact population equality. With a clear mandate to subdivide counties, reconfigure precincts, and create balanced election districts at all levels, political cartographers were intrigued and alarmed by the partisan consequences of threading boundaries through city neighborhoods. Because a line could go forward one of three ways at every intersection, the number of plausibly acceptable configurations was indefinitely large, if not infinite. Although computers could generate and evaluate thousands of nearly optimal solutions, few politicians were willing to cede one of their most prized prerogatives to a programmer or social scientist. Electronic data processing was essential, but only as a service. Moreover, the newly divided legislature demanded independent, equal access for Democrats and Republicans as well as a forum for bipartisan cooperation. In 1978, with little more than two years to prepare for the remap, the ever pragmatic New York legislature established the Legislative Task Force on Demographic Research and Reapportionment.

As long as each party controls a house of the legislature, the Task Force is rigorously bipartisan. Officially, the group has six "members" and two executive directors.[25] Four of the members are legislators; the other two as well as the two executive directors are nonlegislators. The temporary president of the senate (a Republican) and the speaker of the assembly (a Democrat) each appoint one legislator and one nonlegislator, and the minority leaders of both houses (a Democrat in the senate and a Republican in the assembly) each appoint one of the remaining legislators. In addition, leaders in each of the two houses command the loyalty of one of the executive directors, who supervise the technical staff.

As with most labels devised by politicians, every word in the group's name has meaning. "Legislative" identifies its chief client and political patron, "Task Force" implies a sense of mission and urgency, "Reapportionment" is a euphemism for redistricting, and "Demographic Research" reflects a broader concern with race, ethnicity, gender, and age structure as well as intense interest in the impact of census operations on the New York delegation. In 1980, after the decennial head count called for a loss of five House seats, Task Force analyses were the foundation of an unsuccessful lawsuit to force the Census Bureau to adjust for the undercount.[26]

Getting its geographic database in shape for the post-1980 remap was the group's first task. The previous enumeration created headaches for redistricting offices throughout the country. Census officials had underestimated the demand for the detailed local maps and special tabulations needed to comply with the one person, one vote doctrine, and pressing deadlines forced political cartographers to draw up tentative boundaries, which had to be revised or defended in court when more accurate tabulations arrived.[27] Washington promised a smoother count for 1980, but only if states planned ahead. Although the Census Bureau had improved the electronic street maps used to generate rapid, accurate counts for blocks and census tracts, ill-defined boundaries thwarted equivalent tabulations for precincts and other small election districts. Congress responded in late 1975 with Public Law 94-171, which required the bureau to accommodate all states submitting an accurate, detailed description of voting districts by April 1, 1977. Thirty-three states missed the deadline, many because they could not locate reliable maps or promptly realign boundaries with visible features. New York and four other states chose a more expensive solution: contracting with the bureau for block-level counts for areas lacking block data. With detailed data on order at a cost of several hundred thousand dollars, the Task Force set about buying computers, mastering the software, and entering information on party affiliation.

By all accounts, the bipartisan remap was enormously successful, at least for incumbents poised to benefit from a split legislature. With technical support from the Task Force, the presiding members of the two houses worked out a cozy deal in which the assembly sacrificed senate Democrats while the senate wrote off assembly Republicans.[28] (Deals are possible, if not essential, because the head of each house must approve the plan for the other house.) Although the Justice Department rejected the maps for both houses—in their rush to please favored incumbents, mapmakers had failed minority voters—Task Force computers afforded rapid readjustments that won preclearance without forgoing partisan prerogatives. Reinforced by customized constituencies, the split persisted: between 1980 and 1990, the Democratic share of assembly seats advanced from 57 to 63 percent, while Republican strength in the senate held steady at 58 percent.[29] And because successful partisan gerrymandering depends on packing as well as cracking, quite a few assembly Republicans and senate Democrats held equally safe seats. It's hardly surprising that New York incumbents seeking reelection in 1990 posted a success rate (98.9 percent) among the highest in the nation.[30]

New York prides itself on open government—open to legislators and their staff as well as to the general public. In the early 1990s, with collusive bipartisanship well established, the Task Force provided the Democratic and Republican conferences in both houses with individual workstations: separate computers with identical software and data so that party officials could work in private day or night.[31] Outsiders could access the legislature's census and voting data by buying paper maps at $2 per sheet, floppy diskettes at $2 each, or reels of nine-track computer tape at $40—decent prices for anyone not intimidated by the 137 maps, 118 diskettes, or 64 computer tapes required for statewide coverage.[32] Although the public could buy data for individual counties, which would soon remap their own legislatures, anyone seriously interested in statewide redistricting would need statewide coverage. As the Task Force's fourteen-page public access flyer noted, "If you change one district, it affects all those around it."[33]

Invited to participate, black and Hispanic groups presented alternative plans with odd-shaped districts. While African American groups focused on shoring up the four existing black-majority districts in New York City, the Puerto Rican Legal Defense and Education Fund argued that the Voting Rights Act demanded three Hispanic-majority districts, an increase of two.[34] Putting their own interests first, pro-Hispanic cartographers treated African Americans and whites as filler people, while their pro-black counterparts conveniently incorporated clusters of pro-Democratic Hispanic voters. Of course the assembly Democrats and the senate Republicans devised their own plans, predictably at variance with each other as well as with the minority-group remaps.[35]

Little time remained for reasoned debate. Aware that a fight over congressional boundaries might stymie their cozy compromise on assembly and senatorial districts, the legislature had agreed to redraw its own borders first, before taking up the more contentious congressional remap. With the filing deadline approaching and no compromise in sight, ethnic advocates and disadvantaged incumbents appealed to the courts. Compounding the controversy were a governor who vowed to veto any plan with fewer than three Hispanic districts, a state court that crafted a generally pro-incumbent plan with only two Hispanic districts, and a federal court that threatened to impose its own, less incumbent friendly map with three Hispanic districts unless the state produced an acceptable map by July 8.[36] Eager to beat the clock and cut their losses, legislators and the governor accepted the state court's plan.[37]

There's a better way, but most states ignore it. Like New York, they treat political cartography the same as any other legislative chore. Typically a standing or special purpose committee draws up an initial plan, which members then debate, amend, vote on, and if necessary, force into law over the governor's veto.[38] Legislative staff members handle technical details, individual legislators exert clout or whine to the media, joint conferences resolve conflicts between the lower and upper houses, and the presiding officers expedite the process by limiting debate and calling for a vote. Dickering and collusion are rampant, charges of unfairness dominate the headlines, and the aggrieved parties (political or otherwise) file lawsuits or threaten to. When forced to intervene, judges often draft the final map themselves by appointing and directing a presumably neutral special master who understands the state's geography and politics.

The better way is an independent redistricting commission with a

Commissions with initial responsibility for . . .

Congressional and legislative districts

Legislative districts only

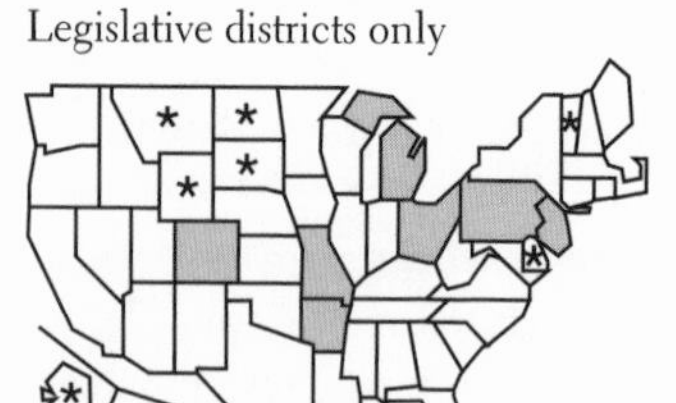

Commissions that back up a deadlocked state legislature

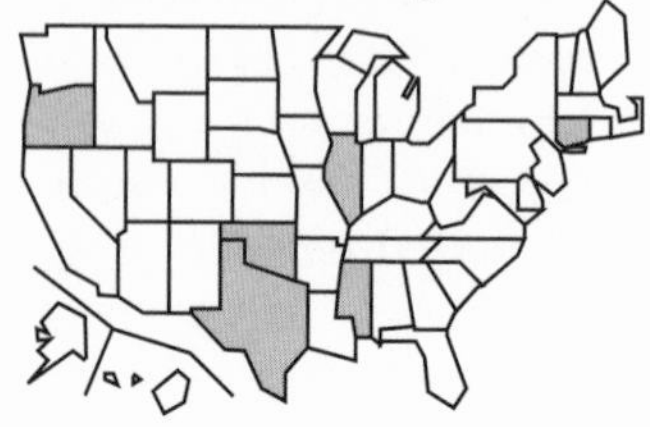

Commissions with a tie-breaking chairperson

Figure 7.3. Roles and responsibilities of independent redistricting commissions. Compiled from Jeffrey C. Kubin, "The Case for Redistricting Commissions," *Texas Law Review* 75 (1997): 837-72.

tie-breaking vote.[39] Legislators set the redistricting criteria themselves but turn the task of redrawing boundaries over to a small group of mapmakers committed to following the criteria consciously and fairly. Commissions typically include legislators or former legislators as well as nonpoliticians. The most efficient formula seems to be a bipartisan commission of five to eleven members, with equal numbers of Democrats and Republicans and a tie-breaking chairperson elected by the other members. Although it's impossible to eliminate politics altogether, removing the remap from conventional legislative channels not only promotes balance and fairness but avoids the delays or outrageous solutions that invite judicial intervention.

Seventeen states have redistricting commissions, but as figure 7.3 reveals, responsibilities vary. Eleven states have commissions that draw up initial plans for legislative districts, and in four of these the commission also reconfigures the state's congressional districts.[40] An additional six states have backup commissions, called into action only if legislators cannot agree on a plan by the constitutional or statutory deadline. Only six commissions have a tie-breaking chair, and these six cut across all three groups.

Although commission-crafted remaps are not immune to lawsuits, states with initial redistricting commissions have largely avoided the contentious, high-profile litigation examined in earlier chapters. Indeed, for the post-1990 remap, Hawaii, Iowa, Montana, and Washington escaped legal challenges to both sets of boundaries, congressional and legislative.[41] Among the seven states that restrict their commissions to legislative districts, only Ohio had its post-1990 plan overturned. According to commission advocate Jeffrey Kubin, Ohio's unconstitutional remap reflected an unusually partisan commission consisting of the governor, state auditor, secretary of state, and two political appointees.[42] A flawed commission structure can undermine an otherwise honorable strategy, and as Kubin concedes, bipartisan commissions are susceptible to bipartisan gerrymanders.

No redistricting commission is as independent as Iowa's Legislative Service Bureau, which provides research and bill-writing support to the legislature.[43] In 1980 the legislature authorized the bureau to redraw the state's legislative and congressional boundaries. By law, redistricting must address four criteria, with population equality foremost, followed by contiguity, the integrity of counties and cities, and compactness.[44] In addition, the bureau may not favor incumbents or consider voter registration or past voting behavior. After examining an array of computer-drawn plans—bureau staff generated 106 alternatives for the post-1990 congressional remap—the director chooses one and presents it to the legislature's Temporary Redistricting Advisory Committee, which holds hearings throughout the state and recommends acceptance or rejection. The legislature cannot amend the maps and must accept or overrule all three plans as a group. If the legislature rejects the package, the bureau presents a second set of plans, and the legislature votes again. Only after a third rejection can the legislature amend the boundaries or draw its own maps.

Iowa's arm's-length redistricting works. In May 1991 the lower and upper houses approved the bureau's boundaries by votes of 93-7 and 39-10, respectively, and the state became the first in the country to complete its congressional remap.[45] Some incumbents were disappointed, but most legislators preferred to take their chances with the nonpartisan remap. Neither party gained a distinct advantage, and as Democratic congressman Dave Nagle opined, "You can do better; you can do worse."[46]

Independent commissions are more widespread than America's meager experience might suggest. Canada, for example, has used provincial electoral boundary commissions since 1964, and Britain has had four nonpartisan reconfigurations since 1948. Both countries sought a peaceful alternative to a self-serving remap by whatever party happened to be in power. Controversies, heated at times, indicate that neutrality remains an elusive goal.

Like the United States, Canada reallocates seats and readjusts boundaries for Parliament's lower house, the House of Commons, every ten years. Each province has a three-member boundary commission, chaired by a judge.[47] Although the commissions are intended to be nonpartisan and independent, selecting the people who draw the lines is unavoidably political, with the chief justice of the Provincial Supreme Court appointing the chair and the Speaker of the House of Commons naming the two commissioners. Unlike American congressional districts, Canada's "ridings" need not be rigidly equal in population. Even so, Canadian courts have coaxed the commissions toward ever greater population equality, and in the "redistribution" following the 1991 census, only 52 percent of ridings departed from the target, "quota" population by more than 5 percent—down from 64 percent in the post-1981 remap.[48] Uneven growth demanded substantial reconfiguration, and many members of Parliament, distressed by the size and shape of their new constituencies, protested vigorously during public hearings in 1994. Angry and frustrated, a majority of MPs voted to suspend boundary adjustment and rethink the use of commissions. But the Canadian Senate blocked the move, and the new House boundaries, with some adjustments after the hearings, took effect for the 1997 elections.[49]

As Britain's experience demonstrates, shrewd political leaders can sway an intentionally neutral process. Like their Canadian counterparts, the commissions that periodically reconfigure counties in England and Wales, regions in Scotland, and boroughs in London must present provisional maps for public consultation.[50] Open comment is a key part of the process, and the ethos of the independent boundary commission includes making adjustments that seem reasonable and warranted. During the mid-1990s readjustment, Britain's Labour Party apparently presented stronger, better-documented arguments than its Conservative rivals. At least that's the conclusion of political geographer Ron Johnston, who studied the hypothetical effects of the 1995 boundaries on the 1992 general election. According to Johnston and his colleagues, the final boundaries would have given Labour five more seats than the provisional boundaries.[51] Although the Conservatives captured a majority of the seats anyway, their majority was smaller—and ultimately more vulnerable—than if the Labourites had been less vigilant and assertive.[52]

Because commission appointments and public comment are unavoidably political, an independent redistricting committee can never be totally neutral. And like numerical constraints on compactness and other allegedly neutral criteria, neutral procedures like a boundary commission inevitably shift power from one party to another.[53] Despite these drawbacks, the redistricting commission remains an attractive alternative for reformers wary about backroom deals and partisan slugfests. And for frustrated yet skeptical voters, handing redistricting over to a few less ostensibly biased political cartographers seems less risky than the far-reaching electoral reforms discussed in chapter 10.

8 *What a Friend We Have in GIS*

COMPUTER-AIDED REDISTRICTING IS A CLASSIC example of technology's unintended consequences. In the early 1960s, political scientists newly aware of the computer's prowess in running population counts and measuring compactness forecast an end to gerrymandering. Columbia University professor William Vickrey, who noted that "whenever the drawing up of the boundaries is left even slightly to the discretion of an interested body, considerable latitude is left for the exercise of the art," proposed "an automatic and impersonal procedure...that will produce results not markedly inferior to those which would be arrived at by a genuinely disinterested commission."[1] Also touting the advantages of eliminating "human discretion," the University of California's Curtis Harris proposed a "possible scientific or impartial method" yielding "a unique solution...with nearly equal populations [based on] a precise rule for compactness."[2] In a similar vein, operations research experts James Weaver and Sidney Hess described an optimizing redistricting method with which legislatures might "avoid compromises and delays" and courts could "avoid partisan pressures and criticism."[3] No one apparently foresaw the computer's role in crafting outrageous racial gerrymanders.

If machines write history, as a technological determinist might argue, it's far from clear which of two competing scenarios is inevitable: intricate gerrymanders or fully automatic redistricting.[4] A Supreme Court convinced that appearances matter is hardly likely to approve severely contorted districts, however purely partisan or race-neutral. Equally clear is a widespread reluctance to embrace wholly automatic redistricting. Although a timely resolution of this thorny task is appealing, politicians are unlikely to entrust the job to computers, especially when the mythical best remap is hidden in a fog of nearly-as-good suboptimal solutions with markedly different effects on influential incumbents. In addition to exploring the lure and limits of optimization, this chapter examines the role of geographic information systems (GIS) in making political cartographers efficiently aware of choices and consequences.

Pervasive use of computers in political redistricting is largely a result of how the Census Bureau collects its data. An accurate census requires a careful preenumeration inventory of households and addresses, which in turn demands a detailed description of city streets and country roads. For the 1970 enumeration, the bureau implemented an electronic street map designed to replace door-to-door enumeration with a mail-out, mail-back questionnaire. In addition to providing printed maps for census field offices and follow-up enumerators, the computer can match a household's filled-out questionnaire, identified by address (e.g., 134 Grafton Avenue), with the appropriate range of addresses on the odd- or even-numbered side of the block (e.g., 100-154 Grafton Avenue). The system's heart, if not its brain, is the DIME file, short for Dual Independent Map Encoding. By linking city blocks to street segments and intersections, DIME principles afford automatic block-by-block population counts, which are then aggregated upward by block group, census tract, city, county, and state. Although concern for privacy limits the kind of information reported at the block level, DIME files support dynamic point-and-click street maps as well as rapid population counts for tentative

Street grid

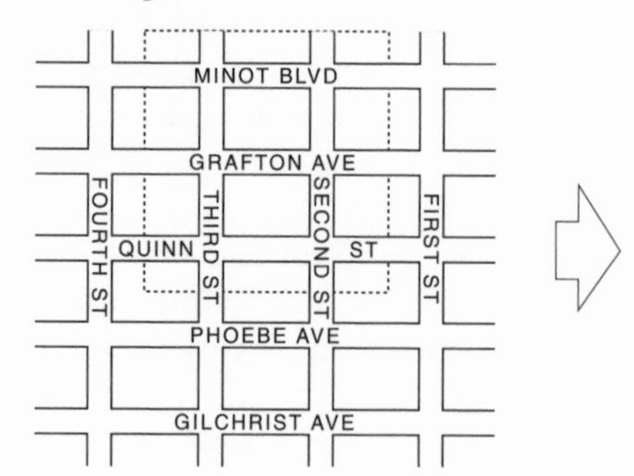

Topological cells

2	1	2	1	2
1	0	1	0	1
2	1	2	1	2
1	0	1	0	1
2	1	2	1	2
1	0	1	0	1
2	1	2	1	2

Figure 8.1. In topology, a street segment is a 1-cell bounded by two 0-cells (intersections or nodes) and cobounded by two 2-cells (blocks).

political districts constructed interactively by adding and subtracting individual blocks—a political cartographer's dream tool.

Mathematician James Corbett, the inventor of DIME, gave the computer eyes for reading and manipulating maps.[5] Corbett's field was topology, a branch of mathematics similar to geometry but concerned with adjacency, not distance. In topological terms, a block (a four-sided *polygon*) is a *2-cell* because it has two dimensions, whereas an intersection point (*node*) is a *0-cell* and a street segment (*edge*) is a *1-cell.* As figure 8.1 illustrates, a 1-cell is thus *bounded* by two 0-cells and *cobounded* by two 2-cells. This "dual" relationship recognizes the street segment as a link between intersections and blocks. If the computer knows the nodes and blocks that bound and cobound each street segment, it can determine which streets surround a specific block or figure out a route from one address to another.

Figure 8.2 shows how a DIME file helps census takers and political cartographers. The file itself is a list of computer records describing the bound-cobound relationships for all links in the urban street network. All nodes, street segments, and blocks are numbered. Adding street names and the ranges of odd and even addresses allows the computer to relate households to street segments, street segments to blocks, and blocks to nodes. Addresses increasing from one end of the block to the other also impart a direction that allows a distinction between left- and right-hand blocks. A separate table holds each node's point coordinates, and another table contains each block's block-group and census-tract numbers as well as its assignment to a voting precinct, a ward, and other political aggregations. The computer uses the point coordinates to plot streets, display data for blocks,

Numbered nodes and blocks

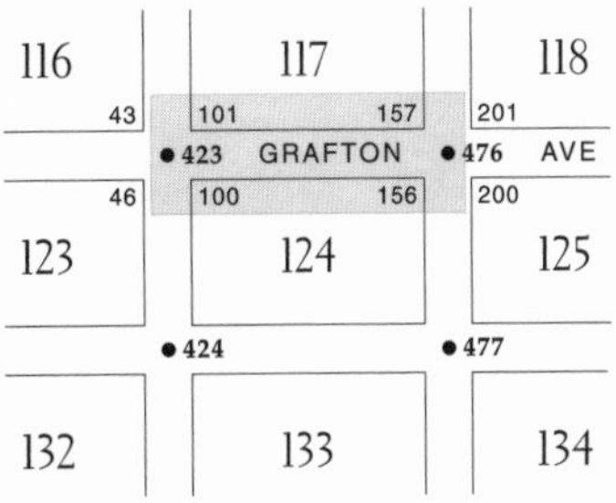

Example of DIME record

STREET:	GRAFTON AVE
ADDRESSES	
Odd range:	101 - 157
Even range:	100 - 156
NODES	
Low end:	423
High end:	476
BLOCKS	
Left side:	117
Right side:	124

Figure 8.2. Individual records in a DIME file relate addresses to street segments as well as describe the topological links among nodes, segments, and blocks.

and respond to queries about a specific block identified on the screen with cursor and mouse. If the user reassigns the block to a new district, the computer can quickly recompute the district's total and minority populations. The computer can also check the file for inconsistencies and pinpoint misnumbered nodes or blocks.

For the 1990 census, the Census Bureau introduced a more powerful and precise database called TIGER, for Topologically Integrated Geographic Encoding and Referencing.[6] A refinement of the bureau's comparatively spartan DIME files, TIGER files provide curved boundaries representing winding streets, streams, railways, and other nonstreet boundary segments. In addition to providing more reliable maps, the new database extended coverage beyond the fringe of metropolitan cities. Available for the 1990 enumeration and remap, TIGER files were the first comprehensive database providing street-level detail for the entire United States.

Unlike the interactive software with which North Carolina and Texas appeased the Justice Department in the 1990s, the earliest redistricting applications promoted compact districts and intact county boundaries. In the 1960s and early 1970s, graphic CRT (cathode ray tube) displays were small, expensive, and rare, and most computer programs ran in "batch mode" on a large mainframe system. With a single computer serving several state agencies or an entire university, each user's "job" awaited its turn in a queue that favored shorter, more urgent "submissions." Despite a turnaround of a day or more for low-priority jobs, operations research experts seeking an optimum solution were rarely upset once they got the bugs out of the program and cleaned up their data. Focused on one or a very small number of "best" solutions, these political cartographers had neither the need nor the opportunity for point-and-click, block-level tinkering.

Optimization software depends on an algorithm, or set of rules, telling the computer what's permissible, what's a good solution, how to make it better, and when to stop. The fundamental goal is to maximize or minimize the value of an *objective function,* such as the disparity between the most and least populous districts, subject to constraints on contiguity, compactness, split counties, and whatever else the analyst chooses to restrict. Each feasible solution yields a value for the objective function, with which the computer evaluates and compares sets of boundaries. The process begins with an initial solution that satisfies the constraints and attempts to improve the value of the objective function by trying out slightly different configurations of district boundaries. If a change improves the solution, it's kept; otherwise the computer remembers the best solution thus far while exploring other possibilities. The machine knows that its algorithm has "converged to an optimum solution" when a further round of trial-and-error iterations fails to improve the objective function.

A simple example shows that the underlying principles are less intimidating than the jargon. The diagram in the lower left corner of figures 8.3 through 8.5 describes a small city that needs to divide its twelve blocks into two districts as nearly equal in population as possible. The street grid is rectangular, and the numbers are the block populations, which sum to 1,000. The objective function is the difference between district populations, which we seek to minimize, and there are two constraints: blocks must remain intact and districts must be fully contiguous. Unlike bushmanders in North Carolina and Texas, parts of a district may not connect only at a point.

Figure 8.3 describes the iterations based on an initial solution that splits the city into two comparatively compact districts. With 385 and 615 residents in the upper and lower districts, respectively, the initial disparity is 230 people. The border is three blocks long, and during the first round of iterations the algorithm examines each of the three street segments in turn, swapping in both directions across the boundary. (The diagram presents swaps in pairs, one above the other, and a thin line marks the boundary segment across which swapping occurs.) The first swap moves a block with 35 residents into the lower district, where population would fall to 350 as the disparity rises to 300—hardly a step in the right direction. By contrast, the other swap along the first block moves 89 people into the upper district, where population would jump to 474 as the disparity drops to 52. Because no

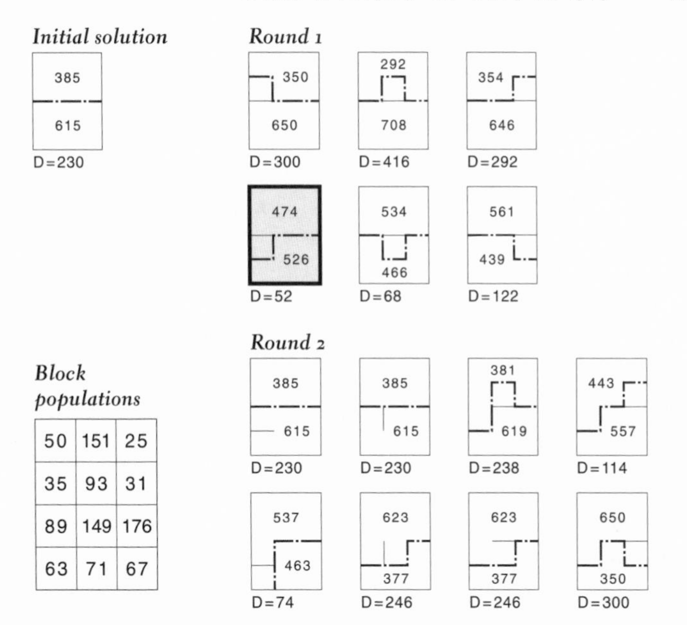

Figure 8.3. Two rounds of iterative trials for a compact initial feasible solution. In this example, round 2 does not improve upon round 1. Diagram in lower left (also used in figs. 8.4 and 8.5) represents a city with a rectangular street grid, twelve blocks, a thousand residents, and substantial variation in population density.

other swap along the initial border yields a greater improvement, this adjustment becomes the starting point for a second round of iterations, along a slightly revised border four blocks long. Because none of the eight swaps along this new boundary lowers the disparity, the optimum solution (highlighted by gray shading and a heavy outline) is the pair of moderately compact districts identified early in round one.

As Figure 8.4 demonstrates, a different initial solution might yield a better optimum. In this example a diagonal, zigzag border five blocks long begins the first round of iterations with a disparity of 52. As in the previous example, the iterations move along the boundary from left to right. And as before, the first pair of swaps yields a measurable improvement, which further trials and a second round of iterations cannot surpass. In the jargon of operations research, the first set of iterations found—and was unable to escape—a "local optimum." This situation is analogous to a severely nearsighted miniature person trying to find the deepest pothole in a generally level street by always moving downhill; whenever this hapless searcher steps into a depression the search is over, even though the street might have at least one deeper pothole.

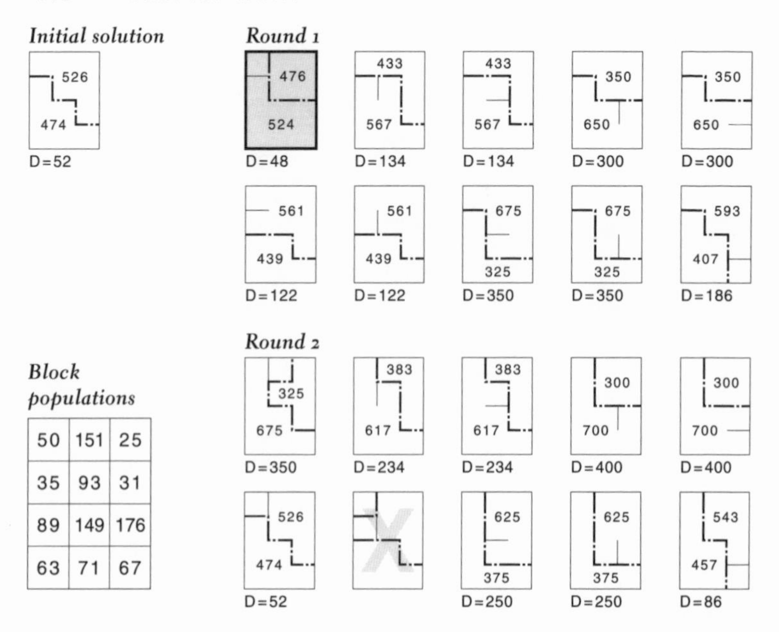

50	151	25
35	93	31
89	149	176
63	71	67

Figure 8.4. Iterative trials based on a measureably better initial solution yield a slight further improvement in round 1. The trial solution marked X is unacceptable because it splits the lower district.

Figure 8.5 illustrates that where you start can make an enormous difference. In this instance the first round of trial swaps lowers the disparity from 402 to 100, and a second round of iterations reduces it further—to zero. Because the disparity can go no lower, the algorithm has found the "global optimum," measurably better than all other, merely locally optimal solutions. Although other initial solutions might have led to the same boundary, these three examples demonstrate that a good initial solution is no guarantee of finding the global optimum.

For a state with multiple districts and hundreds of thousands of blocks, a swap-and-test algorithm will probably miss the global optimum. The number of feasible boundary configurations is so large that not even the fastest computer can generate and test them all. To boost the odds of finding a solution not much worse than the global optimum, a competent algorithm will generate at random a diverse set of initial feasible solutions, track down an optimum for each, and save the best result. Unless the search stumbles on a solution like figure 8.5, the objective function's theoretical optimum, the computer that cannot test all possible boundary configurations has no way of knowing

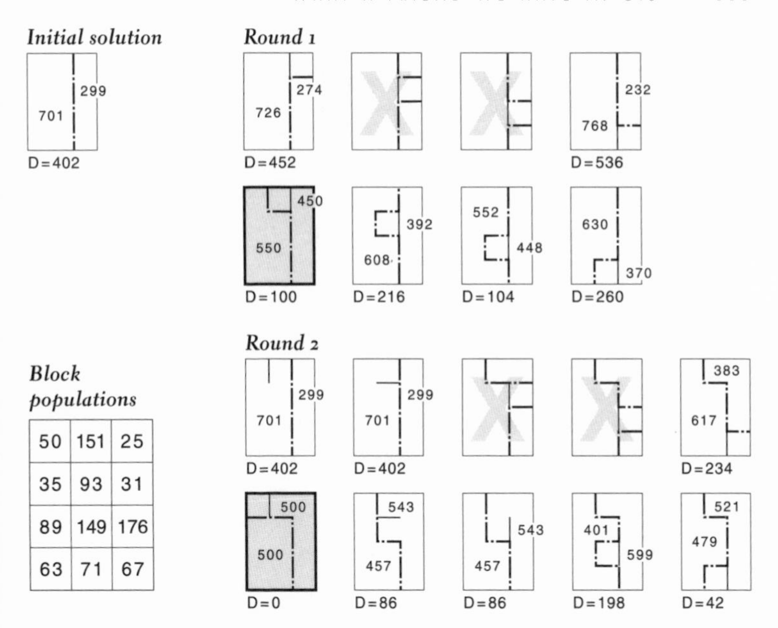

50	151	25
35	93	31
89	149	176
63	71	67

Figure 8.5. Two rounds of iterative trials based on a measurably worse initial solution find a global optimum with districts exactly equal in population. Trial solutions marked X violate the contiguity constraint by splitting the rightmost district.

whether it has found the global optimum. And even if the search uncovers a theoretical optimum like perfectly balanced populations, there is no way of knowing whether other equally optimal solutions exist.

To increase the likelihood of finding the global optimum, an analyst might beef up the algorithm with "simulated annealing."[7] In the steel industry, annealing is a process of allowing molten metal to cool slowly so that molecules form a hard yet malleable material that resists cracking; by contrast, rapidly cooled steel is brittle and less pliant. A redistricting algorithm simulates annealing by injecting random changes that temporarily make the solution worse in order to slow down an otherwise "greedy" algorithm's rapid convergence on a convenient but markedly suboptimal local optimum like those in figures 8.3 and 8.4, where the search was trapped early in round one. Even if the improvement is not dramatic, simulated annealing at least ensures a more careful exploration of the broad range of feasible solutions.

Another strategy for overcoming the myopia of swap-and-test algorithms borrows a technique from large discount retailers that depend on strategically located warehouses. Geographers call this approach "capacity-constrained location-allocation modeling" because the computer not only locates two or more supply centers with fixed capacities but also specifies the source and amount of goods shipped to each demand center.[8] While the primary goal is to solve the "transportation problem" by finding allocations that minimize shipping costs from warehouses to stores, the objective function and constraints might also consider spoilage, the costs of land and utilities, and the inherent inefficiency of very small and very large warehouses.

Redistricting has analogous goals and constraints: city blocks and rural precincts are the demand points; each representative's single, centrally located district office is a supply point; each block is assigned to a district office; and assignments that minimize travel time to district offices and lessen population disparities among districts promote representative democracy. This analogy was attractive to redistricting experts, who could adapt existing location-allocation software. Although legislators and representatives need not have a single, centrally located district office, a solution that minimizes travel time to a hypothetical central location yields arguably compact districts.

Compactness and population equality are key goals in James Weaver and Sidney Hess's classic proposal for automated redistricting, published in 1963 in the *Yale Law Journal*.[9] Figure 8.6, a condensed version of their flow chart, describes an iterative algorithm based on the transportation problem but modified to minimize the sum of squared distances between the centers of election districts (EDs), where people live and vote, and the centers of legislative districts (LDs), which serve multiple EDs. Squaring distance focuses the algorithm's attention on remote EDs, which can greatly reduce a district's compactness. After accepting ED coordinates and population data, the algorithm enters coordinates for an initial set of LD centers: good guesses that the analyst hopes might lead to an optimal solution. For each iteration, the process computes distances and finds a solution that minimizes total "transport cost," computed as the sum of squared distances all people would travel between the centers of their election and legislative districts. Because Weaver and Hess used the transportation problem's capacity constraints to equalize LD populations, the resulting allocations usually split at least a few EDs among two or more LDs. And because EDs must remain intact, the algorithm reassigns each split ED to the LD allocated the largest share of its population. After computing LD populations as well as dispersion statistics for the overall solution, the algorithm checks whether the

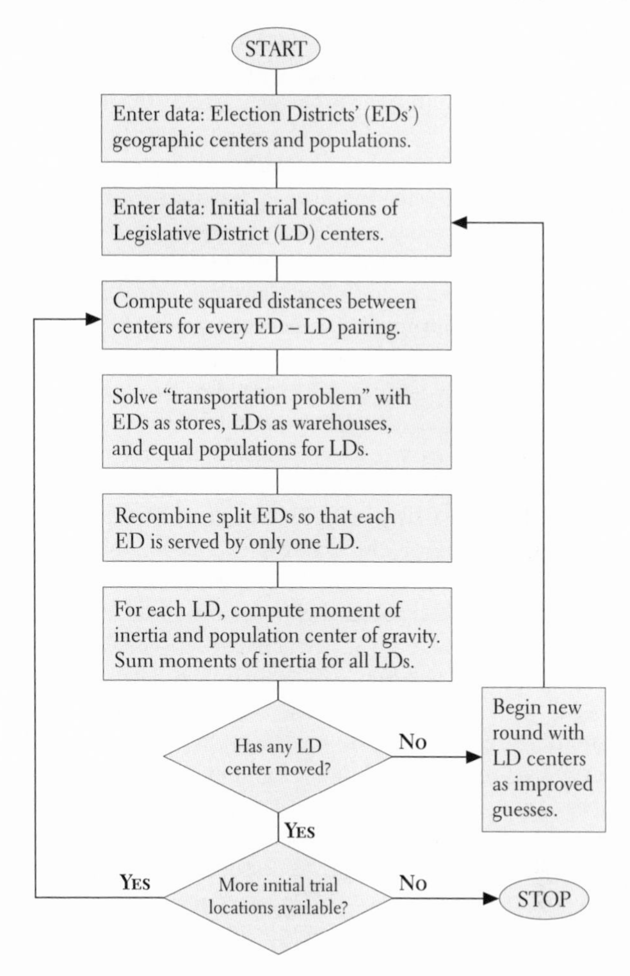

Figure 8.6. Flow chart describing the location-allocation redistricting algorithm of Weaver and Hess. Redrawn and condensed from drawing in James B. Weaver and Sidney W. Hess, "A Procedure for Nonpartisan Districting: Development of Computer Techniques," *Yale Law Journal* 73 (1963): 288-308; diagram on 303.

LD centers have moved from their locations at the beginning of the iteration. If so, the new centers become the good guesses for another round of iterations. If not, the algorithm either halts or begins anew with another set of initial trial locations. A variety of initial solutions are necessary because location-allocation algorithms, like their track-and-swap counterparts, are easily seduced by local optima.

If you sense that the process is not as fully automated and straightforward as its advocates imply, you're right. In addition to elusive global optima, redistricting officials and their consultants must sort out various ways of framing the problem as well as the inherent trade-offs between population equality and compactness. Because the ban on split EDs overrides the equal-population constraint used in solving the transportation problem, the Weaver-Hess algorithm favors population-weighted compactness over an obsessive pursuit of population equality.[10] An alternative strategy might tolerate a 1 percent disparity among LD populations when the algorithm solves the transportation problem; this compromise is especially appropriate if the EDs are blocks rather than census tracts, which have much larger populations and might easily be split in urban settings. Among other variations, using linear distance rather then squared distance avoids undue influence by remote EDs with small populations, and replacing straight-line distance with highway distance or travel time recognizes that voters, unlike the fictive flying crow, rarely travel in straight lines. Indeed, with geographically precise TIGER data, a highly competent computer could estimate road distance or travel time as well as orchestrate a hybrid, hierarchical process that uses location-allocation modeling to generate an optimal set of districts based on census tracts and track-and-swap optimization to fine-tune their boundaries at the block level.

Is automated redistricting worth the effort? University of Washington geographer Richard Morrill thought so. In 1972, after state legislators failed to agree on a redistricting plan, a federal court appointed Morrill its special master and gave him a month to partition the state into forty-nine legislative districts with no more than a 1 percent disparity in population.[11] (Because each district elected one senator and two representatives, a single plan was sufficient.) In addition to prohibiting contact with politicians and the media, the court imposed several other constraints: respect natural barriers like Puget Sound and the Cascade Divide; keep Indian reservations and Seattle's black community intact; promote compactness; and avoid splitting counties, cities, and census tracts. With too little time to develop a database and acquire software, Morrill relied largely on outline maps, census data, the competing plans drawn up by the Democratic Senate and the Republican House, and a keen sense of fairness, efficiency, and the state's geography. His map ("the court plan") generated some controversy but was accepted without undue rancor by the Republicans, who saw in it a slight advantage, as well as by the Democrats, who griped more loudly yet captured both houses in the 1972 elections. Morrill also wrote the court plan for the state's seven congressional districts, each a cluster of seven legislative districts.

Curious whether a computer might have drawn better boundaries—and if so, what they would look like—Morrill revisited the problem three years later with a location-allocation model that approximated the court's requirements.[12] Instead of relying on straight-line distances, as in the Weaver-Hess model, he treated EDs and LDs as nodes in a transport system and adopted a network-based algorithm, which provided separate solutions for three kinds of distance: straight-line, highway, and travel time.

How Morrill used the computer is revealing. Conscious of the overwhelming computational demands of partitioning several hundred election districts into forty-nine legislative districts, he first established optimal boundaries for the seven congressional districts and then used separate models to divide each congressional district into seven legislative districts. To promote efficient convergence to an optimum, he referred to the 1972 map and placed initial centers at either the congressional district's geometric center or the node with the largest population. And to satisfy the court's mandate that two large Indian reservations and Seattle's black neighborhoods remain intact, he drastically reduced distances among election districts within these communities. No further adjustments of the data were needed because clusters of urban election districts tended to preserve the integrity of cities, and the road-distance and travel-time models were unlikely to ignore Puget Sound and the Cascades.

Results were impressive. Although Morrill's court plan was not far off, the computer generated measurably more compact districts. How much better is apparent in the aggregate road distances calculated for each plan in thousands of miles (table 8.1).[13]

Although the statewide totals were roughly similar for the court plan and the straight-line solution, the computer's effort to minimize road distance produced a 3 percent improvement. And when programmed to minimize travel time, the computer beat Morrill's manual solution by 2 percent for road mileage and 5 percent for travel time. Equally impressive are the reduced travel times for six of the seven districts (table 8.2).[14]

These gains might seem minimal, but Morrill, who had studied the state's road network and population closely, was ecstatic: "The generally high quality and overall superiority of the computer version was, to tell the truth, unexpected, indicating either that we were unable to discover the more efficient patterns manually or that we intentionally

Table 8.1 Aggregate road distances for Morrill's original court plan and three redistricting plans constructed several years later by computer

District	Court plan	Straight line	Road distance	Travel time
1	2,525	2,258	2,292	2,421
2	11,354	14,928	13,483	11,354
3	20,942	18,620	18,274	19,660
4	39,832	39,970	38,675	38,675
5	22,419	22,232	20,756	20,756
6	5,303	4,287	4,934	6,099
7	3,014	3,504	3,482	3,415
Total	105,389	105,304	101,896	102,380

Table 8.2 Morrill's computer-generated travel-time plan yielded more compact configurations for six of the seven districts

District	Court plan	Travel time	Improvement
1	1,048	1,005	4.1%
2	4,770	4,394	7.9%
3	6,986	6,275	10.2%
4	8,669	7,973	8.0%
5	9,692	9,970	-2.9%
6	2,190	2,119	3.2%
7	1,478	1,447	2.1%
Total	34,883	33,183	4.7%

or unintentionally emphasized other criteria such as the preservation of incumbents by trying not to change the districts too much."[15]

Comparing aggregate distances and travel times with the four maps in figure 8.7 indicates that these improvements largely reflect adjustments to boundaries in rural, sparsely inhabited areas. In the Seattle-Tacoma area, for instance, Districts 1, 6, and 7 (not labeled on the maps) reflect inherently compact groupings of urban residents that, if delineated at all, have relatively little effect on overall compactness—at least as measured by road distance or travel time. Real gains require reconfigured boundaries in comparatively remote regions, where many constituents are well removed from their district's hypothetical center. Even so, the impressive 8 and 10 percent improvements to Districts 2 and 3 involve exchanges with their metropolitan neighbors. Equally revealing is the trade-off between Districts 4 and 5, where an 8 percent gain in the time-distance compactness for District 4 requires a 3 percent loss in District 5. To maintain the integrity of the regionally important Tri-Cities (Richland-Pasco-Ke-

1972 Congressional districts: "the court plan"

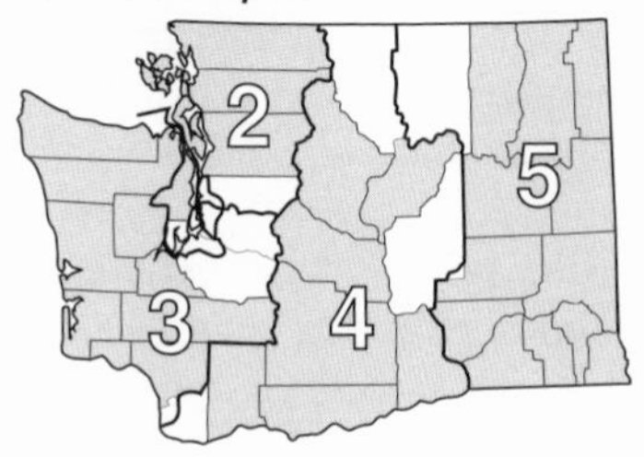

Boundaries optimized for straight-line distance

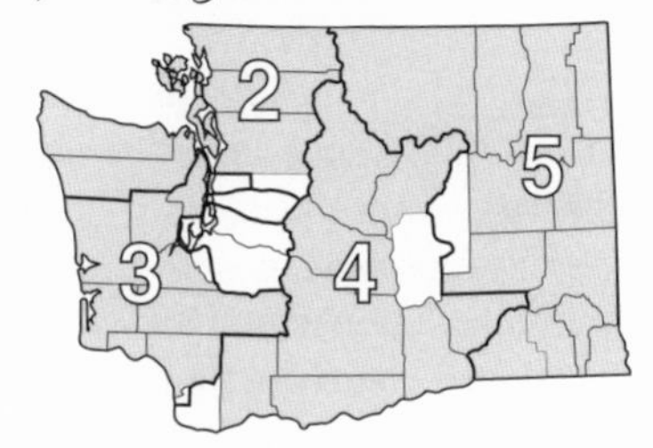

Boundaries optimized for road distance

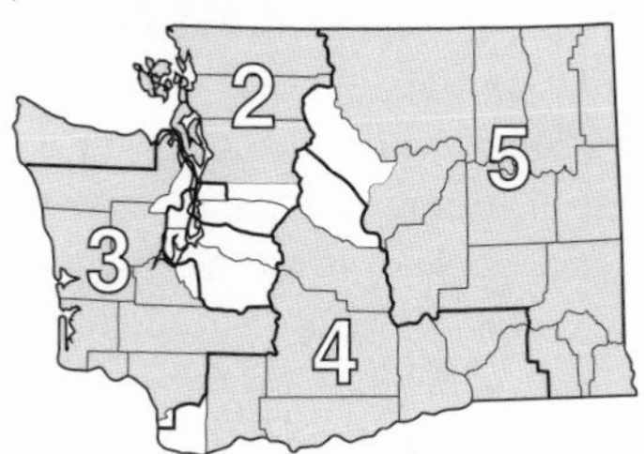

Boundaries optimized for travel time

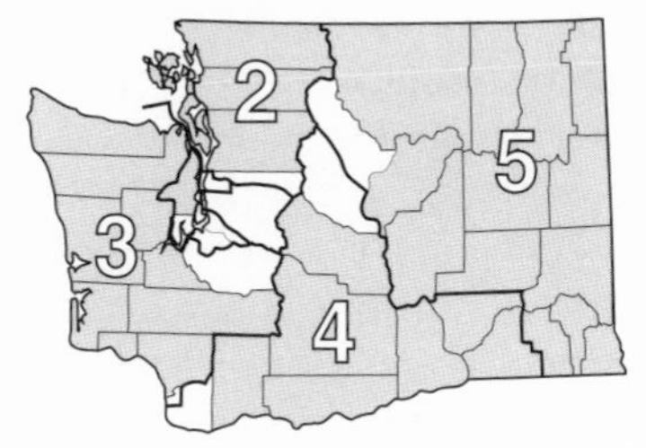

Figure 8.7. Comparison of the state of Washington's "court plan" congressional boundaries with the optimal boundaries geographer Richard Morrill generated by computer several years later. Compiled from various maps in Richard L. Morrill, "Redistricting Revisited," *Annals of the Association of American Geographers* 66 (1976): 548-56.

newick) conurbation in southeastern Washington, the computer reassigned a large portion of the court's Fourth District to its eastern neighbor. In applauding this exchange, Morrill also commended the computer solution for improved compactness in Districts 3, 4, and 5 as well as improved regional identity in Districts 4, 5, and 7.

Morrill's enthusiastic endorsement did not herald widespread automated redistricting. Although computers plotted maps, calculated populations, and helped legislators manually explore the effects of swapping counties or census tracts, only two states, Delaware and Iowa in the late 1960s, ever submitted to algorithmic optimization.[16] Or so the literature suggests—sketchy details imply that these experiments, like Morrill's, were little more than academic exercises.[17] This dearth of serious trials is especially significant for the 1990s, when a flurry of law review essays touted algorithmic remapping as nonpartisan and objective.[18] Despite the obvious attractions of optimization strategies for maximizing population equality, enhancing minority representation, or promoting compactness, neither the federal courts nor the Department of Justice pressured state legislatures toward full automation. The technology was available, but likely customers weren't buying.[19]

Political scientist Micah Altman explains this resistance. Although computers are becoming ever faster, political redistricting is "a computationally hard problem...widely believed to be computationally intractable."[20] For the huge number of city blocks and rural census units involved, he argues, the fastest computer with the most efficient algorithm offers little more than a good guess: a very good guess, perhaps, but not necessarily the global optimum. Moreover, the problem can be structured in various ways, depending on how the algorithm addresses the competing goals of population equality and compactness, which can be addressed either in the objective function or as constraints; on how the computer measures compactness; and on how the process treats minority groups, if at all. This variation affords a diverse suite of solutions, undermining claims that automated, algorithmic redistricting is neutral, nonpartisan, and inherently objective. Because party leaders, judges, or Justice Department officials can choose the solution they like, manipulation is inevitable. What's more, proposals to legislate the computational strategy in advance are suspect because well-heeled stakeholders with consultants or their own software can pierce the "veil of ignorance" believed to make automated districting fair and unbiased.[21] As Altman warns, "automation may not eliminate the opportunity to manipulate politically, but instead shift that opportunity toward those groups that have access to the most extensive computing facilities and expertise."[22]

Algorithmic approaches have another drawback: the potential embarrassment of formally stated goals. How would it look, for instance, if a redistricting committee overtly constrained its solution to protect some incumbents and target others? Interactive mapping seems a more subtle way to exploit the power of computer graphics, electronic census data, and the battalion of consultants eager to advise legislatures, political parties, and minority advocates.[23] And if state law bars redistricting officials from putting incumbents' addresses or voting data in the computer, party officials and civil rights proponents can set up their own systems and generate alternative plans with which to lobby legislators, Justice officials, or federal judges. As the transcripts of several key cases reveal, maps drawn up by these unofficial political cartographers led to some of the past decade's most infamous gerrymanders.

In Georgia, for instance, Kathleen Wilde, an attorney for the American Civil Liberties Union, drew up the "Max-Black Plan," which demonstrated the feasibility of three minority-majority congressional districts encompassing all substantial concentrations of African Americans. (A four-district plan was apparently tried but found impossible.) As an advocate for the legislature's black caucus, Wilde was (in the court's words) "in constant contact" with the Justice Department's Civil Rights Division, which refused to approve the state's first two submissions.[24] Among other evidence of an unconstitutional racial motive, the federal district court noted "countless communications, including notes, maps, and charts, by phone, mail and facsimile between Wilde and the DOJ team." Informed by Wilde's insights, Justice officials offered detailed advice on how the state might comply. Georgia officials soon realized that the preclearance mavens would not approve any map with fewer than three minority districts.

A key witness was Linda Meggers, director of reapportionment services for the Georgia legislature and (in the court's opinion) "probably the single most knowledgeable person available on the subject of Georgia redistricting."[25] Meggers, who worked closely with legislative leaders, testified that her colleagues had no intention of discriminating against black voters. Indeed, their first submission included two minority-majority districts and one minority-influence district. Although their second submission raised the latter district's black voting-age population from 35 to 45 percent, Washington's well-briefed watchdogs held out for the 52 percent majority in the third submission.

Meggers's testimony included a demonstration of the interactive computer she used to make maps and tally percentages. The judge who wrote the majority opinion was highly impressed by the "extremely sophisticated" computer's prowess in drawing maps "at the census block level with the greatest of ease."[26] He also applauded the computer's pet name, Herschel: "Only Georgians truly understand the depth of respect that is accorded to this equipment by such an appellation."[27] (The judge was clearly a fan of famous University of Georgia running back Herschel Walker.)

Interactive racial gerrymandering was equally apparent in Texas, where a software package named REDAPPL (for Redistricting Application) helped legislators construct the three highly intricate minority-majority districts that the Supreme Court roundly condemned in *Bush v. Vera*. As in the Georgia case, the district court hearing initial arguments was impressed by the computer's ability to work at block level with racial and ethnic data: "If the Legislature intended to allocate voters on the basis of race, REDAPPL certainly provided a readily available, efficient means of doing so."[28] What's more, the slightest shift in a boundary registered instantly as a change in black and Hispanic voting clout. As Chris Sharman, the legislature's cartographic technician, told the court, "The problem is when you draw on this computer, it tells you the population data, racial data. Every time you make a move, it tabulates right there on the screen. You can't ignore it."[29]

Federal judges in North Carolina took a less cynical view of interactive redistricting. Ordered by the Supreme Court in *Shaw I* to reexamine the issue of illegal racial gerrymandering, the district court endorsed the state's new map, including the infamous I-85 District, as a "narrowly tailored" solution that served the "compelling interests" of geography and the Voting Rights Act.[30] In the majority opinion, what Justice Sandra Day O'Connor had denounced as bizarre became the inevitable consequence of the *Gingles* decision, modern computing, a demographic database with 229,000 census blocks, and an ambitious legislative staff, which added precinct election results from several recent statewide elections: "This not only made possible a new degree of refinement in getting the numbers right (including the rather incredible achievement of mathematically perfect equal population districts), it also encouraged the drawing of boundary lines with more obvious irregularities and facial oddities than typically occurred under the less sophisticated methodology formerly available."[31]

Has GIS eroded the significance of shape? So it seems. In *Shaw II*, the Supreme Court struck down North Carolina's racially motivated gerrymander and affirmed the principle, clarified earlier in *Miller v. Johnson*, that the constitutional issue is race, not geometry. And in *Bush v. Vera*, a Texas case in which the high court overturned three oddly shaped racial gerrymanders, tacit approval of equally contorted but purely partisan gerrymanders may have established a "technological momentum" defense that forever undermines the importance of compactness, at least at the federal level. Where race is not an issue, it's apparently up to state legislatures and state courts to ward off GIS-inspired assaults on political aesthetics.

9 *A Tale of Two Censuses*

THE 1990S CLOSED WITH A NEW REDISTRICTING controversy: a heated debate over Census Bureau plans to use statistical sampling in its year 2000 enumeration. On one side were the bureau's brass, the Clinton administration, most Democrats, and the American Statistical Association; on the other side were an assortment of traditionalists and most Republicans. Proponents argued that statistical adjustment for nonresponse would yield a more accurate count, while opponents contended that the Constitution demanded an "actual enumeration."[1] The Supreme Court, which might have resolved the dispute, ruled that the government could, if it wanted, report two sets of results: a traditional head count for reapportioning the House of Representatives and a statistically adjusted tabulation for apportioning federal dollars and redrawing electoral boundaries.[2] Paying for a two-count census was another matter: in spring 1999 an impasse between the White House and the Republican-controlled Congress threatened to shut down the government, as had happened in 1995 when a presidential veto of a balanced-budget bill kept "nonessential" federal workers off the job for six days.[3] Fearing public reaction to a similar mess, GOP lawmakers agreed to fund the dual-number census.

Republicans had two reasons to fear statistical estimation: the undercount and the overcount. Uncounted persons, who are frequently nonwhite, homeless, or illegal aliens, typically live in less affluent districts represented by Democrats.[4] Successful corrections for the undercount would tend to make these districts smaller, thereby shifting Democratic voters into nearby districts, which might be Republican. Similarly, persons counted more than once often have more than one residence, typically in areas represented by Republicans. Correcting for the resulting overcount would tend to lower the populations of these affluent districts, which in turn would expand their borders and add voters who might be Democrats.

How severely these adjustments would affect individual districts is anyone's guess. But consider the plight of the Republicans in figure 9.1, showing a hypothetical two-district state in which registered Democrats enjoy a clear 60/40 percent majority in District A, while registered Republicans cling to a narrow 50.8/49.2 percent edge in District B. To simplify the numbers, I've set the resident and voting populations of both districts at 1,000 and 600, respectively. The upper-right example describes the discovery of a whopping 20 percent undercount in District A: a disparity that triggers the transfer of one-twelfth of A's population and voters to B. Note that although adjustment has increased the overall population, the number of voters remains unchanged: just because the Census Bureau thinks an additional 200 persons are living within District A's old boundaries doesn't mean they'll suddenly show up at the polls. Even so, transferring one-twelfth of District A's territory, population, and 60/40 Democratic voters wipes out the Republicans' slim margin in District B.

The lower-left example in figure 9.1 describes the milder but nonetheless threatening consequences of uncovering 100 double-counted residents who don't belong in District B. The equal-population requirement demands that District A give up one-twentieth of its resident and voting populations to District B, thereby lowering the Republicans' edge to 50.3/49.7 percent. And as the lower-right ex-

Initial districts, no adjustment

1,000 residents
600 voters
360 Democrats
240 Republicans

Democrats in District A have a 60% - 40% advantage.

1,000 residents
600 voters
295 Democrats
305 Republicans

Republicans in District B have a 10-vote edge.

Adjustment for 200 undercount in A

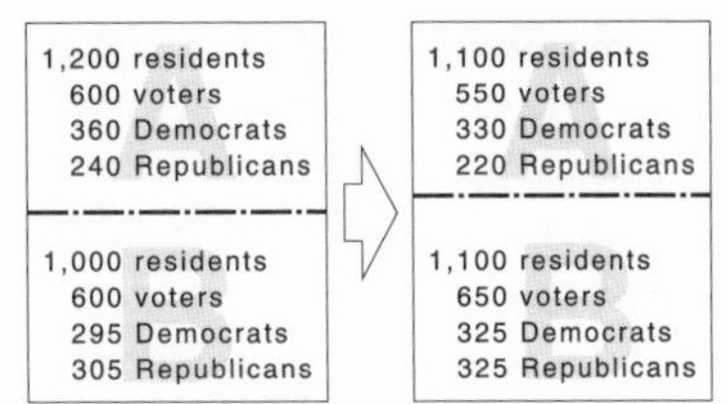

Adjustment for 100 overcount in B

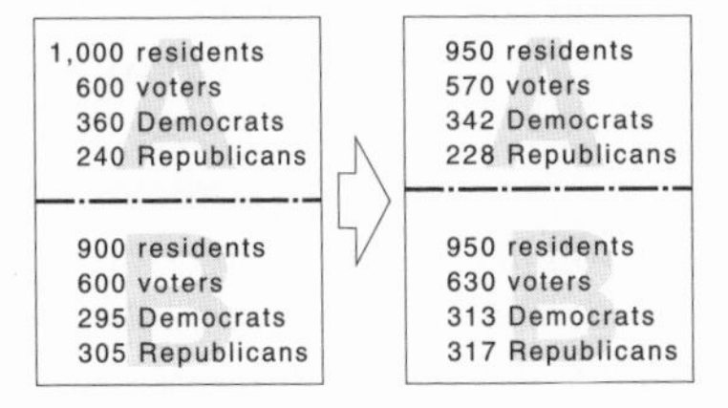

Combined adjustments

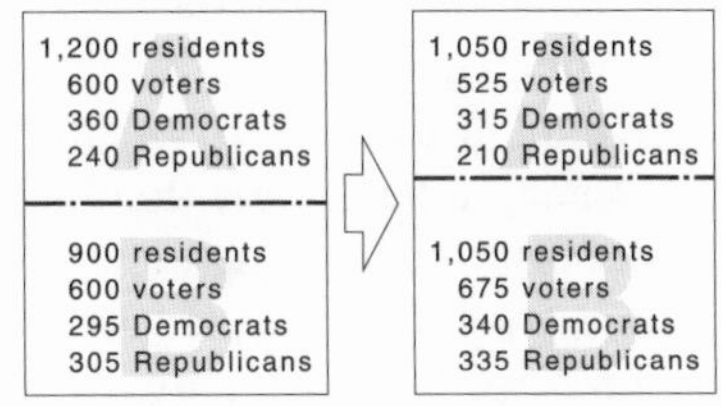

Figure 9.1. Consequences of adjusting for an undercount of two hundred and an overcount of one hundred in a hypothetical two-district state in which Democrats have a strong majority in one district while Republicans hold a narrow lead in the other. Adjustment to equalize population moves the boundary between Districts A and B and transfers voters as well as residents. Transferred voters are assumed to reflect average voting behavior in District A before adjustment.

ample illustrates, the combined adjustments for a 200-person undercount in District A and a 100-person overcount in District B transfer one-eighth of A's population and voters to B, giving the Democrats a slight advantage—unless the Republicans can gerrymander their supporters into a more favorable constituency.[5]

Although I cooked the numbers to illustrate a process, the net undercount of 100 in my hypothetical example is not as outrageous as it might seem. A net error of 100 persons for an estimated population of 2,100 is 4.8 percent: well above the estimated 1.3 percent net nationwide undercount of nonblacks for 1990 but below the 5.7 percent rate for the black population.[6] Applying the overall national net undercount of 1.6 percent to the numbers in figure 9.1 would have done little damage to my hypothetical Republicans, except in a push comes to shove year when every vote really does count. Even so, the wide disparity between the black and nonblack rates—social scientists call this the "differential undercount"—combined with the tendency of black voters to favor Democrats suggests that Republican fears are plausible if not well founded.[7] However minor its effects, adjustment clearly won't do the GOP much good.

What worries the Census Bureau is that Americans, especially African Americans, are getting harder to count. Figure 9.2, a time-series plot of the estimated net undercount, mixes good news with bad. The good news is that the most recent census was more accurate than its 1940 and postwar counterparts. The bad news is an alarming rise in the net undercount between 1980 and 1990. What's more, as the U.S. General Accounting Office points out, the estimated net error of 1.6 percent for 1990 does not mean that over 98 percent of the population was counted accurately. Indeed, the net undercount of 4.7 million persons shown in the graph conceals a gross error of approximately 16 million persons, computed by adding an estimated overcount of 6 million to an undercount of 10 million.[8]

Demographers seem confounded by the overcount, which includes pets and other fictitious persons, infants born after the April 1 enumeration date, people with dual residences, and college students counted both at their parents' homes and at their dormitories.[9] A clearer picture emerges for the racial, gender, and geographic differences that contribute to the undercount. In 1990, for instance, blacks were more than four times as likely to be missed as nonblacks, males were four times as likely to be missed as females, and black males were fourteen times as likely to be missed as white females.[10] Renters, foreign-born citizens, and illegal aliens were particularly evasive. According to former Census Bureau director Martha Farnsworth Riche, "The Hispanic male renter aged 18-34 is the hardest to count."[11] It's hardly surprising that a map of the net undercount (fig. 9.3) shows markedly higher rates in California, New York, and most of the South and Southwest—areas with substantial numbers of blacks, Hispanics, and other minorities. These are the states most likely to benefit from corrective sampling.

Understanding the need for correction requires an appreciation of the enormous challenge of counting everyone at home just once. Because the Census Bureau depends heavily on a mail-out, mail-back questionnaire (in 1990, letter carriers delivered a self-administered mail-back survey to 84 percent of all households, while census enumerators dropped off a similar survey at another 10 percent, principally in rural areas)[12], accuracy begins with a reliable list of household addresses and conscientious follow-up for nonresponse. In compiling and cross-checking its "address frame," the bureau merges its own list from the previous census with more or less up-to-date information from the Postal Service and offers local governments an opportunity to suggest corrections.[13] Workers at over four hundred field offices

Net undercount: millions of persons missed

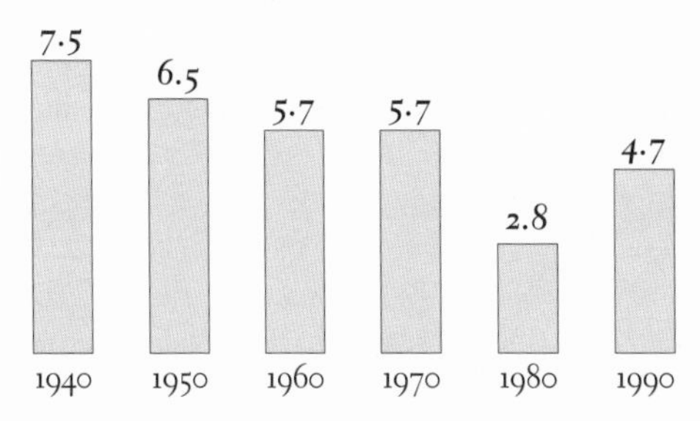

Figure 9.2. Net undercount, 1940-90. Compiled from U.S. General Accounting Office, *2000 Census: Progress Made on Design, but Risks Remain,* report GGD-97-142 (Washington, D.C., July 1997), 7.

Net undercount rate, as percentage of estimated population, by state, 1990 census

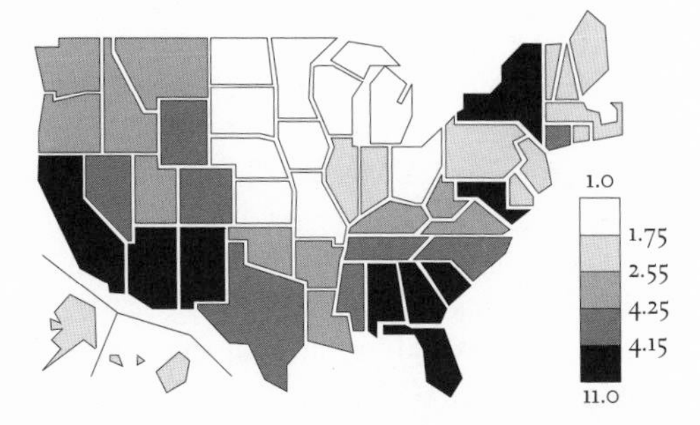

Figure 9.3. Net undercount by state, 1990. Compiled from Panel on Census Requirements in the Year 2000 and Beyond, Committee on National Statistics, Commission on Behavioral and Social Sciences and Education, National Research Council, *Modernizing the U.S. Census* (Washington, D.C.: National Academy Press, 1995), map on 36.

then go door-to-door to determine whether each address represents only one household. Field personnel also note vacant or abandoned buildings as well as locations where structures have been torn down or are under construction. Verifying address lists and following up for nonresponse required a huge workforce: the Census Bureau hired 501,000 temporary workers to plan and carry out the 1990 enumeration, and Census 2000 triggered a similar recruiting effort.[14] To help them get around, the bureau maintains a huge set of detailed street-level base maps and supplies enumerators with a quarter million copies.[15]

Detailed maps and aerial photographs are especially important in the preenumeration canvass of remote areas, Indian reservations, and other places where people pick up their mail at a post office. Enumerators required to drop off a mail-back questionnaire or retrieve a mailed-out census form need to plan their routes in advance and know where to try again if no one answers. On the Menominee Indian Reservation in rural Wisconsin, for instance, census takers must count nearly 5,000 people, many living in trailers scattered across 235,000 acres of forest. Instead of addresses, dwellings have descriptions like "yellow trailer with brown trim" or "house at the second fence on highway Double A."[16] If a resident answers the door, occupancy is obvious. If not, the census worker must walk around the outside and peer in windows. Because tribal members are suspicious of government snoops, census officials hire Native Americans to verify occupancy and count heads. The Menominee Reservation is one of three sites at which the Census Bureau tested its methods with a "dress rehearsal" two years before the year 2000 enumeration.[17]

Counting the homeless is especially problematic. In 1990 the Census Bureau compiled lists of shelters in cities with 50,000 or more residents and sent 23,000 workers into the streets, parks, shelters, bus stations, missions, YMCAs, all-night movies, and flophouses to count people who might otherwise evade enumeration.[18] Ordered to estimate age, race, and sex without disturbing anyone's sleep, they usually worked in teams of two or three. One shift counted shelter occupants from 6:00 P.M.to midnight, and a second shift tallied people on the street between 2:00 and 4:00 A.M. and persons entering or leaving boarded-up buildings from 4:00 to 8:00 A.M. All together the "S-Night" (for shelters and street) sweep counted 179,000 persons in shelters and approximately 50,000 living in alleys, parks, tunnels, bus stations, and abandoned structures. Although the list of shelters was judged fairly accurate, the inexperienced enumerators largely ignored dumpsters, bridges, sewers, and other comparatively secure locations favored by the more wary street people. Homeless advocates and social scientists roundly denounced the flawed effort, which probably missed more than half the homeless population.[19] Martha Burt, who directed a 1987 Urban Institute study that put the number at 500,000 to 600,000, preferred comparatively stress-free daytime counts at soup kitchens. "A lot of people avoid shelters," she noted, "but they will eat."[20]

However elusive, homeless persons were far less troublesome than the 34 million households that lost or ignored their census forms, costing taxpayers an additional $560 million for follow-up.[21] Despite a barrage of television ads and other costly outreach and promotional efforts, the mail-response rate of 65 percent in 1990 had dropped alarmingly from 75 percent in 1980 and 78 percent in 1970.[22] Calling, visiting, and otherwise persuading nonrespondents to participate had driven the cost of counting heads upward, in constant 1990 dollars, from $11 per housing unit in 1970 to $20 in 1980 and $25 in 1990. If the downward trend in participation continued, census officials feared a 55 percent nonresponse rate in 2000 would generate a follow-up workload of 53 million cases and raise the total tab, in unadjusted dollars, from $2.6 billion in 1990 to nearly $5 billion in 2000. Sampling, they argued, would save money as well as provide a more accurate count.

The Census Bureau's original sampling plan promised to cut costs as much as $800 million by estimating the size and composition of the hardest-to-enumerate households.[23] To ensure geographic consistency, follow-up efforts would raise the response rate to 90 percent in all of the nation's 60,000 census tracts. (Intended for social science and demographic studies, *census tracts* are largely homogeneous areas averaging roughly 4,000 people and 1,700 housing units.) If 80 percent of a tract's households responded, enumerators would need to visit only one of every two remaining households. If only 50 percent of households responded, visits to four out of five nonresponding units would boost the response rate to 90 percent. Estimating the characteristics of the remaining 10 percent by statistical extrapolation would save time and money, and a nationwide postcensus quality-control sample of 750,000 households would allow statistical correction of likely errors in the direct count.[24]

What seemed logical to statisticians and social scientists intimidated congressional Republicans. House Speaker Newt Gingrich called sampling "a dagger aimed at the heart of the Republican majority" and vowed to fight the plan in court as well as in budget votes.[25] Where statisticians saw increased accuracy, the GOP saw potential fraud. "The notion of statistical sampling scares the living daylights out of me," said Republican redistricting expert Benjamin Ginsberg. "It is so subject to political manipulation."[26] Alarmed by the Census Bureau's sampling plans, Republican leaders threatened to withhold funding unless the Democrats agreed to scrap sampling.[27] In November 1997, after President Clinton threatened to veto a key appropriations bill, the Republicans and the administration agreed to compromise: while House lawyers sought a court ruling on the constitutionality of statistical estimation, census officials would develop plans for two different censuses, one with traditional direct enumeration and the other with sampling to correct for nonresponse.[28]

Under the compromise, the Census Bureau tested sampling the following April in dress rehearsals in Sacramento, California, and in Menominee County, Wisconsin, which includes the Menominee Indian Reservation.[29] Glitches were rare, and bureau officials pronounced the trials a success. Even the low mail-back rates—54 percent in Sacramento, 41 percent on the reservation—were slightly better than expected.[30] Although conscientious follow-up picked up most of the remaining households, the direct counts fell well below their statistical counterparts. In Sacramento the traditional census counted 349,197 persons, while sampling put the total at 403,313, a figure supported by a state estimate of 392,834, based on growth projections, drivers' licenses, and utility hookups.[31] In Menominee County, direct enumeration counted 4 percent fewer people than statistical estimation. No one knew the exact population for certain, but the 4,779 residents estimated with sampling was much closer to the true figure than the 4,595 people counted by direct enumeration.[32] In different ways, of course, both numbers had a use.

Could politicians learn to live with the ambiguity of a dual head count? Better get used to it, the Supreme Court warned. On January 25, 1999, the justices announced a 5-4 decision with plums for both political parties.[33] Republicans were heartened to hear that the Census Act, as amended in 1976, "prohibits the proposed uses of statistical sampling in calculating the population for purposes of apportionment."[34] Democrats were equally pleased to learn that the sampling was acceptable for other uses of the census, including redistricting and allocating federal funds. Writing for the majority, Justice Sandra Day O'Connor observed that the 1976 amendments "required that sampling be used for such purposes 'if feasible.'"[35] In recognizing that the census has diverse uses, the high court noted that the law not only endorsed the "mandatory use of sampling in collecting non-apportionment information" but gave the secretary of commerce (who oversees the Census Bureau) "substantial authority to determine the manner in which the decennial census is conducted."[36]

Debate quickly shifted from law to money. With Democrats now supporting a two-number census, Republicans offered to pay whatever the Census Bureau needed to ensure the accuracy of a traditional, one-number enumeration. No thanks, the Clinton administration responded. Blaming the high court for a projected $2 billion cost increase, census officials quickly unveiled a new two-track strategy.[37]

Track one, providing reapportionment data, included a costly but thorough canvass, on foot or by phone, of all nonresponding households. Where repeated efforts failed and a residence was clearly occupied, the bureau's "automated data processing system" would "account for missing data" by "applying statistical techniques."[38] The 1990 census had used a similar strategy for including known households that census takers were unable to contact after attempting three visits and three telephone calls at different times on different days.[39] Track two, serving nonapportionment needs, would correct track one's "raw data" using a "scientific sample of approximately 300,000 housing units [and] regional grouping."[40] Because this second strategy was essentially what the Republicans had feared all along, GOP leaders threatened to both starve and micromanage census officials into submission.[41] One proposal, ostensibly to increase accuracy, encouraged local governments to challenge figures and demand a recount.[42]

Few aspects of the sampling controversy are as puzzling as the equanimity with which politicians and census officials ignore a form of computerized adjustment called "imputation," which the Census Bureau has used for decades. An automated editing process, imputation involves separate operations known as "substitution" and "allocation."[43] Substitution is necessary because some households, despite census takers' best efforts to contact and cajole, simply will not fill out and return their questionnaires. Allocation is essential because some people return the survey without answering one or more questions or write in "Don't know." In the first instance, the bureau adjusts for nonresponse by "substituting" the complete set of answers from a similar household. To help the computer find a logical match, the enumerator questions neighbors, building managers, or letter carriers about the number of residents and their ages, genders, and other characteristics.[44] And in the second instance the bureau's computer "allocates" the missing responses from the census form of a household with similar answers to a few related questions.[45] Don't be appalled: imputation might be fudging, but it's unavoidable. Without substitution, census counts would be unrealistically low, and without allocation, inconsistencies in the data would undermine demographic research and policy planning.

Substitution is far less common than allocation. In 1990, for instance, substitution generated data for 0.6 percent of the population while allocation supplied one or more missing items for another 16.2

Persons substituted for noninterview, percentage of population, 1990 census

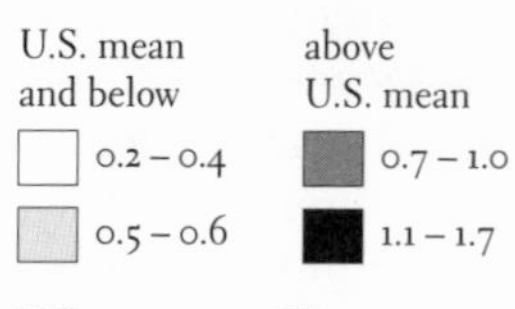

U.S. mean = 0.6%

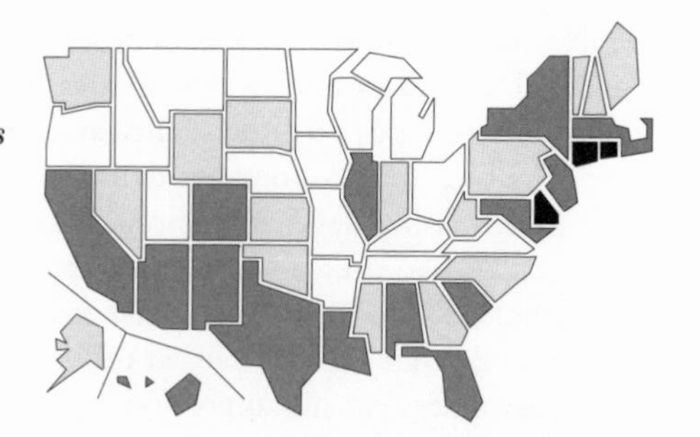

Persons with one or more items allocated, percentage of population, 1990 census

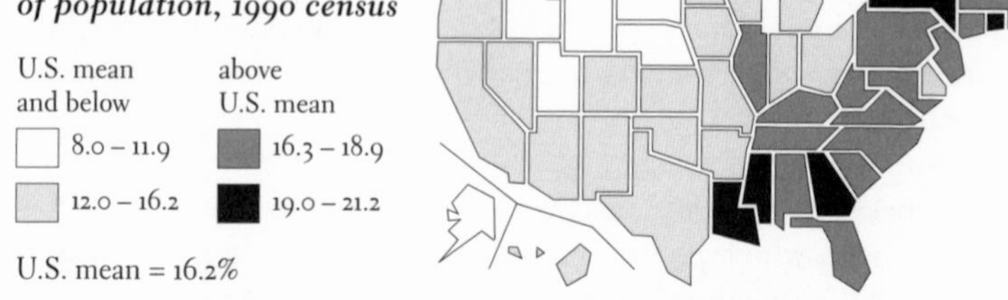

Figure 9.4. Rates of substitution (above) and allocation (below) were generally higher in states with substantial minority and foreign-born populations. Compiled from data in U.S. Bureau of the Census, *1990 Census of Population: General Population Characteristics, United States,* report CP-1-1 (Washington, D.C., November 1992), 589.

percent.[46] Although the Census Law requires complete and truthful cooperation, the Census Bureau rarely prosecutes violators—imagine the cost if it did. An accurate census depends on voluntary participation and goodwill. This also explains why the bureau prefers not to advertise imputation: some citizens could lose faith in the census, while others might figure Why bother? and not respond.

Despite sensitivity about imputation, census officials are frank about its use. Details are readily available at the back of many published census tabulations, as my maps in figure 9.4 readily attest. The upper map shows above-average substitution rates in Illinois, the metropolitan Northeast, and states bordering Mexico. Statewide rates ranged from 0.2 percent for North Dakota to 1.7 percent for Rhode Island, with average rates generally higher for large cities (1.2 percent) than for rural areas (0.5 percent) and small towns (0.4 percent)—places with comparatively few minority and foreign-born residents.[47] By contrast, the lower map reveals a more marked concentration of allocation in the East and Southeast, perhaps a reflection of poverty and illiteracy. At least it's easy to explain why Mississippi, our poorest and least literate state, combines the highest statewide allocation rate (21.2 percent) with a below-average substitution rate (0.5 percent).

The need for substitution contradicts the alleged purity of an "actual enumeration" and weakens objections to sampling. Counting heads—over a quarter of a billion of them—is a complex process requiring endless quality checks, edits, and adjustments. Census statisticians have used sampling to evaluate coverage since 1950, and they are confident that statistically adjusted estimates are more reliable than traditional tabulations.[48] And what's more efficient, they ask, than incorporating sophisticated postenumeration survey techniques into the process up front, rather than years later when estimating net undercounts from the same information?[49] As an act of faith, they vigorously reject traditionalist assertions that unadjusted counts, with or without substitution, are somehow more real.

At the heart of the matter, I am convinced, is the public's traditional sense of numerical reality. If counting seems at all possible, most of us consider a count more believable than an estimate. We don't want scientists or statisticians guessing at the tab for the groceries in our shopping carts or the gas we pump into our cars, after all. Also, terms like "head count" and slogans like Be Counted reinforce the idea that an accurate "actual enumeration" is possible.[50] But other explanations intrude, including widespread ignorance of mathematical statistics and mistrust of government officials, pollsters, and marketing mavens. Cultural, racial, and class bias also seem relevant, especially among those who believe that people not easily counted—illegals, dimwits, and others "not like us"—are just as well ignored. It's easy to equate persons not counted with people who don't count.

Census 2000 stirred up another controversy. For the first time, Americans could declare more than one race. As figure 9.5 illustrates, the new census form offers fifteen separate choices, including blanks for naming a specific native tribe or "other" race. With this flexibility, golf champion Tiger Woods, who describes his mixed heritage as "Cablinasian," could check boxes for white (*Ca*ucasian), *bl*ack, American *In*dian, and *Asian*.[51] By contrast, the 1990 questionnaire would have forced him to choose just one. According to Census Bureau guidelines, if Woods had written in "Caucasian-Black-Indian-Asian," he would have been recoded as white, whereas the alphabetical equivalent would have made him Asian.[52]

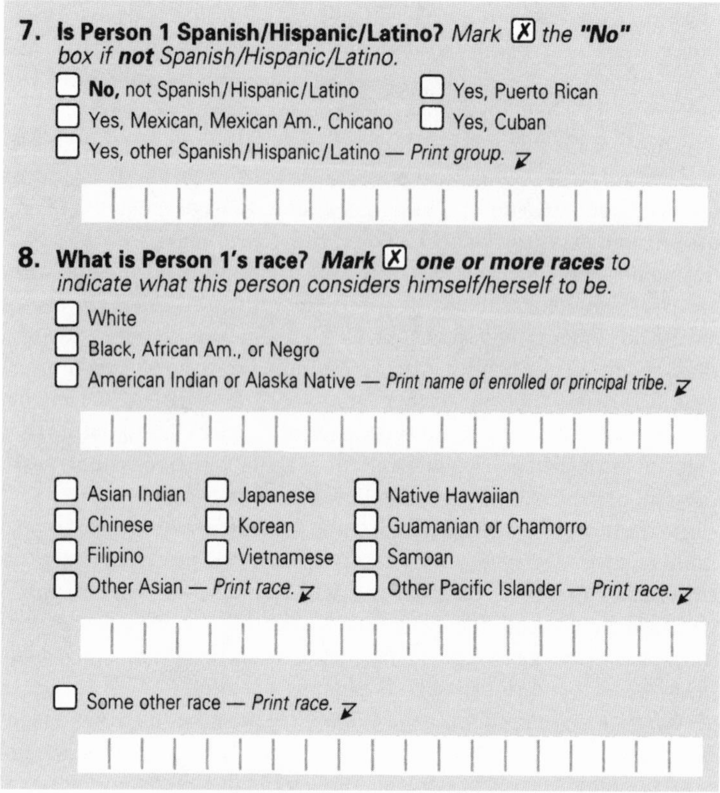

7. Is Person 1 Spanish/Hispanic/Latino? *Mark* ☒ *the* ***"No"*** *box if* ***not*** *Spanish/Hispanic/Latino.*

☐ **No,** not Spanish/Hispanic/Latino ☐ Yes, Puerto Rican
☐ Yes, Mexican, Mexican Am., Chicano ☐ Yes, Cuban
☐ Yes, other Spanish/Hispanic/Latino — *Print group.*

8. What is Person 1's race? ***Mark*** ☒ ***one or more races*** *to indicate what this person considers himself/herself to be.*

☐ White
☐ Black, African Am., or Negro
☐ American Indian or Alaska Native — *Print name of enrolled or principal tribe.*

☐ Asian Indian ☐ Japanese ☐ Native Hawaiian
☐ Chinese ☐ Korean ☐ Guamanian or Chamorro
☐ Filipino ☐ Vietnamese ☐ Samoan
☐ Other Asian — *Print race.* ☐ Other Pacific Islander — *Print race.*

☐ Some other race — *Print race.*

Figure 9.5. Hispanic-origin and race questions on the Census 2000 questionnaire.

Census officials were well aware that their race question was flawed.[53] It not only compelled a person with mixed lineage to choose one parent's race over the other's but ignored the growing number of interracial marriages, which had increased from 0.4 percent of all marriages in 1960 to 2.0 percent in 1990.[54] Moreover, 2.9 percent of all couples (married or otherwise) were interracial, and their households reported 2 million children, up markedly from 500,000 in 1970.[55] In addition to slighting a growing cohort of mixed-race citizens, the 1990 census lumped together Asians and Pacific Islanders eager to identify themselves more narrowly as Korean, Chinese, or Native Hawaiian.[56]

Abandonment of the single-race designation followed four years of study by the Office of Management and Budget, which coordinates statistical information for federal agencies.[57] OMB considered but rejected a vague mixed-race category, which offered simplified tabulation but little (if any) gain in informativeness or reliability. African American advocates were strongly opposed because a "multiracial" category, as an alternative to "black," might dilute their clout at preclearance time.[58] Legal scholar Christine Hickman, who noted that a truly accurate multiracial count might include 30 to 70 percent of all African Americans, most Native Americans, almost all Latinos and Filipinos, and a surprising number of whites, argued for "leav[ing] the multiracial inquiry off the race line and isolating this inquiry on a line of its own."[59] The statistics czars agreed, but their "mark one or more" decision went a step further and kept the word "multiracial" off the census form altogether.

To improve the responsiveness and accuracy of Spanish speakers, the new questionnaire asks about Hispanic origin immediately before inquiring about race. (On the 1990 questionnaire, the race and Hispanic questions were item items 4 and 7, respectively.) The order in which questions are asked is important because race is confusing to many Hispanics, who can be white, black, American Indian, or even Pacific Islander. Pretests of the new form revealed that when the race question included a multiracial category, putting the Hispanic origin question first not only increased the response rate (from 90.0 to 92.5 percent) but upped the preponderance of Hispanics who considered themselves white from 56.1 to 67.5 percent and lowered the proportion reporting "other race" from 24.9 to 15.7 percent.[60] And replacing the "multiracial" category with an invitation to "mark one or more" race categories raised the response rate further (to 93.9 percent).[61] By carefully disentangling ethnicity and race, the Census Bureau enhanced the meaningfulness of both items.

More problematic, of course, are the varied ways of tabulating the data for redistricting. For example, a "max-black" count would tally the number of persons selecting African American, with or without another race, whereas a "min-black" count would exclude persons listing more than one race. In Census Bureau parlance, the first approach produces "all-inclusive" totals, while the second yields "single-race" counts.[62] But these aren't the only strategies. For example, a "historic-series" approach compatible with earlier tabulations assigns multiracial persons to a single minority group if they choose two groups, one of which is white or "other," or places them in a "multiple race" category if they declare two or more minority (nonwhite) groups.[63] And a fractional approach, perhaps fairer, would first divide each person by the number of races declared and then accumulate totals similar to the all-inclusive counts. In this case a person listing both black and white would contribute only one-half to the black count, and Tiger Woods, by listing four races, would contribute only one-fourth. For this reason African American advocates would prefer to base redistricting on all-inclusive counts.

Despite the fifteen separate racial responses on the new questionnaire, five general categories seem sufficient for assessing compliance with the Voting Rights Act.[64] Otherwise, state and local redistricting officials could be overwhelmed by separate block-level counts for individual races as well as a glut of two-, three-, four-, or five-way combinations, each of which must be subdivided twice: once to break out the voting-age population and again to separate out Hispanics. To avoid needless complexity, the Justice Department agreed to a simplified redistricting data set with single-race and all-inclusive counts reported for five racial groups plus one or two additions.[65] Single-race counts, which will not sum to 100 percent without both an "other" and a mixed-race category, require seven groups:

- White alone
- Black or African American alone
- American Indian and Alaska Native alone
- Asian alone
- Native Hawaiian and Other Pacific Islander alone
- Some other race alone
- Two or more races

In contrast, all-inclusive counts, which inherently sum to more than 100 percent, need only six general categories:

- White alone or in combination with one or more other races
- Black or African American alone or in combination with one or more other races
- American Indian and Alaska Native alone or in combination with one or more other races
- Asian alone or in combination with one or more other races
- Native Hawaiian and Other Pacific Islander alone or in combination with one or more other races
- Some other race alone or in combination with one or more other races

Census officials anticipate no noteworthy differences between the all-inclusive and single-race counts for blacks, although significant local disparities could arise.[66] How then would Justice officials assess compliance with the nonretrogression provision (Section 5) of the Voting Rights Act, which seems to require data comparable to the 1990 tabulations? One solution is to look at both numbers—the max-black and min-black counts bracket whatever 1990 counterpart one might construct—and hope that compliance is obvious. If not, the larger, all-inclusive count could lessen the complexity of whatever new boundaries preclearance might require.

Do geographic differences in the net undercount really matter? As my hypothetical two-state example (fig. 9.1) illustrates, the answer is not a straightforward yes or no. Most of the time differences among states will not matter, and it's a tribute to the Census Bureau that they don't. But as chapter 2's discussion of reapportionment formulas revealed, a few thousand more residents might make a big difference, as Massachusetts demonstrated by narrowly losing the 435th House seat in the 103rd Congress. But statistical adjustment is not as simple as urban mayors like to think. Although Boston's minorities might have been underreported, postenumeration sampling suggests that corrections for undercounts and overcounts in 1990 would have affected only six states: California, Georgia, and Montana would each have gained a seat at the expense of Oklahoma, Pennsylvania, and Wisconsin.[67] Gubernatorial hand-wringing notwithstanding, statistical adjustment would not have helped Massachusetts, New Jersey, and New York, meaningless winners of the fictitious 436th, 437th, and 438th seats (table 2.3).

If nothing else, the two-number Census 2000 will demonstrate dramatically that scale matters. Separate but almost always unequal counts for each block and census tract will highlight the greater uncertainty of census enumeration at the local level and confirm the relative stability of congressional districts, with markedly larger populations than their state and municipal legislative counterparts.[68] The two numbers will also stoke fires of discontent, no doubt, as whiney minority advocates discover neighborhoods for which corrections appear inadequate while outraged traditionalists attack census officials for deleting real people from the adjusted count.[69] Few critics, I'll wager, will note that any census—even the most thorough, costly, and precise enumeration—is obsolete by the time the reapportioned and redistricted Congress holds its opening session.

10 *Beyond Boundaries*

REMEMBER LANI GUINIER? IN APRIL 1993, a newly inaugurated President Clinton named her assistant attorney general for civil rights, pending Senate approval. Although Bill and Hillary had known Lani since their Yale Law School days, her appointment was hardly a case of cronyism. Guinier's credentials were impressive: degrees from Radcliffe and Yale, outstanding service awards for four years in the Justice Department's Civil Right Division, seven years of litigation experience with the NAACP Legal Defense and Education Fund, including a successful voting-rights lawsuit against Arkansas while Clinton was governor, and a tenured professorship at the University of Pennsylvania.[1] An articulate scholar, Guinier had published several articles that championed "proportional representation"—dirty words to the *Wall Street Journal,* which condemned the president for appointing a "Quota Queen."[2] Debate focused on a divisive issue: whether African Americans, who constitute an eighth of the population, should occupy a similar proportion of House seats. Had Guinier's critics read her writings carefully (or at all), they would have discovered a cogent case for replacing bushmanders with comparatively compact multimember districts. And had Clinton not withdrawn the nomination before Guinier could defend herself, the American public might have enjoyed an enlightening discussion of a better way to draw political boundaries and elect legislators.[3]

Although pundits and politicians railed at Guinier's alleged attack on American traditions, the *Historical Atlas of United States Congressional Districts,* compiled by geographer Kenneth Martis, reveals ample precedent for districts with two, three, or four representatives.[4] At various times between 1793 and 1843, five states used what Martis calls "plural districts."[5] The first was Massachusetts, which created one two-member and two four-member districts for the Third Congress (1793-95). Although the Bay State tried plural congressional districts only once, Pennsylvania set up a two-member district for the Fourth Congress (1795-97), maintained at least one multimember district through the Twenty-seventh Congress (1841-43), and would have done so much longer had Congress not banned the practice with the Apportionment Act of 1842. The ban also affected Maryland and New York, which had used plural districts since the Eighth and Ninth Congresses, respectively. Inspired by its neighbors, the New Jersey legislature tried it once, by partitioning the state into three two-member districts for the Thirteenth Congress (1813-15), but quickly reverted to statewide at-large elections.

Plural districts were especially suitable for New York and Pennsylvania, where population density varied widely from the remote, rugged Adirondacks and Appalachians to the emerging metropolises of New York City and Philadelphia. Committed to keeping counties intact while giving urban centers a modicum of population equality, legislators had responded to the 1830 census with a two-member district for the city of Philadelphia and a four-member district for Manhattan and the Bronx.[6] In addition, the Empire State designated pairs of counties around Brooklyn, Elmira, Hudson, and Syracuse as two-member districts, while the Keystone State lumped three counties west and south of Philadelphia into a three-member district. Martis, who called the practice "one of the oddities of early state congressional district statutes," noted that winner-take-all balloting typically enabled a single political party to control a district's entire delegation.[7] Besides outlawing plural districts, the Apportionment Act of 1842 banned the even more tyrannical abuse of "general ticket" elections, in which all congressional candidates ran at large on a statewide party ticket.[8]

Lani Guinier was after more than multimember districts. Her proposal called for elections like those in most European countries as well as a handful of American municipalities.[9] Cambridge, Massachusetts, for instance, has used proportional representation since 1941 to elect its nine-member city council and six-member school committee.[10] The city's unusual voting system seems a bit complicated, but residents prefer it. The five times that disgruntled politicians collected sufficient signatures to force a public referendum, Cambridge voters chose to keep proportional representation.

Voting is straightforward. Separate paper ballots for the council and school committee invite voters to rank acceptable candidates in order of preference, starting with 1 for their first choice, 2 for their second, and so forth. A voter can rank one or all of the candidates, or any number in between, but a tied rank invalidates the votes for the similarly ranked candidates.[11] The hardest part is remembering not to rank anyone you wouldn't want in office.

Because voters might be influenced by a candidate's position on the printed ballot, Cambridge promotes fairness by listing candidates' names alphabetically and then rotating the list so that each name appears first on an equal number of ballots.[12] Thus, if twenty citizens run for city council, election officials must print twenty different ballots. Multiple ballots might seem a boon to a lucky local print shop, but the cost is minor. Whatever the price, rotation is more equitable than putting the leading party or candidates whose names start with A at the top or making the election a crap shoot by ordering the candidates randomly.

Counting the ballots is comparatively complicated, at least to the uninitiated. Before 1997, when the city turned to a computer, the count took a committee of tellers five to seven days.[13] To win a seat, a candidate must meet a "quota," calculated by dividing the number of ballots cast by the number of seats plus one, and adding one to the result. The math is simple, really: if 25,000 residents vote in the race for nine council seats, the quota is 2,501. The computer then counts the number of first-choice votes for each candidate, declares the "first-count" winners, whose tallies reach or exceed the quota, and distributes any surplus votes to remaining candidates. For example, if 4,000 voters marked Betsy Witherspoon as their first choice, the resulting 1,499-vote "surplus" is redistributed pro rata to the 4,000 voters' next-choice candidates.[14] The arithmetic might appear a bit daunting, but the system doesn't waste votes. After identifying additional winners and redistributing their surplus votes, the computer eliminates candidates with fewer than fifty votes and reassigns their totals to each voter's next choice. Each additional round eliminates the weakest remaining candidate and redistributes his or her votes. The process continues until all seats are filled.

In the lingo of electoral theorists, Cambridge employs the "single transferable vote" system of proportional representation, a form practiced in Australia, Ireland, and Malta.[15] Among "PR" advocates the method has two other names: "choice voting" and "preference voting."[16] Nearly two dozen American cities have tried choice voting since 1915, when residents of Ashtabula, Ohio, used it to elect a diverse council of merchants and laborers; Democrats, Republicans, and Socialists; Catholics and Protestants; and English, Italians, and Swedes. Ohio was a seedbed for choice voting: Hamilton, Toledo, Cleveland, and Cincinnati once had it, and Cambridge's computer counts ballots according to the "Cincinnati method."[17] Nearly two dozen cities tried it, including Boulder, Kalamazoo, New York, Sacramento, and Wheeling, West Virginia.

According to political scientist Douglas Amy, the complex count was less a factor in its decline than machine politics: proportional representation almost always threatened party bosses, who retaliated with organized campaigns to rewrite the city charter.[18] Another factor was the Cold War: fear of Communism precipitated a rash of repeals in 1947, when New York City voters abandoned the choice-voting system used for city council elections since 1936.[19]

Simple arithmetic explains why political establishments felt threatened. Almost all American cities, counties, and states elect their legislatures by a "first past the post" plurality system in which the candidate with the most votes wins. (Some jurisdictions, typically in the South, require a runoff election for any race in which no candidate receives more than half the votes.) If all council candidates run at large and every voter casts as many votes as there are seats, a patronage machine can consistently run the show. And where election districts are the basis for the traditional American system called "single-member plurality" voting, a savvy machine with a license to gerrymander can capture control with fewer than half the votes.

Machine control is markedly more difficult under proportional representation. In Cambridge, for instance, any group sufficiently numerous and cohesive to rally a tenth of the voters behind a single candidate is assured a seat on the council. In effect, group members create a virtual election district for themselves, unconstrained by where they live. What's more, if the group cannot agree on a single candidate, the next-choice count affords a worry-free way of expressing personal preferences without having to guess the winner. After all, if your first choice is my strong second choice and vice versa, neither of us need fret about wasted votes if our two candidates can collectively "reach quota." Not wasting votes is what proportional representation is really all about.

By any measure, choice voting has served Cambridge well. Turnout for municipal elections is significantly higher than in other eastern Massachusetts cities similar in size and demography.[20] Voters understand the ballot, believe that fairness and responsiveness justify the complex and once-tedious tallying, and are proud that their city has had an African American councillor since the 1950s.[21] Advocates of electoral reform look to Cambridge as a model of progressive vitality and credit proportional representation with fewer attack ads and a fuller discussion of issues.[22]

Although Lani Guinier applauds Cambridge's pioneering effort, her prototype for proportional representation is the simpler "cumulative voting" system used in Chilton County, Alabama.[23] Seven commissioners and seven school board members run at large, and each voter receives seven votes on each ballot, to allocate as he or she pleases. A voter can give one vote each to seven candidates, allot five votes to one and two to another, or "plump" all seven votes on a single favorite. Although Chilton County voters might have more difficulty than their Cambridge counterparts in figuring out a strategy, the county's tellers have the less tedious task of merely adding up each candidate's votes and awarding seats to those with the seven highest totals.

Because it's easy to waste votes on a popular candidate with lots of supporters, electoral reformers consider cumulative voting a form of "semiproportional" representation: more equitable than the single-member plurality but not as fair as choice voting.[24] Hartford, Philadelphia, and Washington, D.C., among other cities, use a third form of proportional representation called "limited voting" because voters in an at-large election cannot cast as many votes as there are seats.[25] In a race for seven seats, for instance, each voter might have only three votes. According to PR experts, limited voting becomes more inclusive—that is, more favorable to a popular minority candidate—as the difference between the number of seats and the number of votes increases.

Chilton County adopted cumulative voting in 1988, after a federal court ruled that countywide at-large elections in Alabama illegally diluted the strength of black voters.[26] Rather than impose remedies, the court encouraged counties to work out a solution with the Alabama Democratic Conference, which had filed the class-action lawsuit. The standard solution, single-member districts, would not work well in largely rural Chilton County: African Americans, who composed just under 12 percent of the population, were so widely scattered that a contiguous, defensibly compact black-majority district was impossible without expanding the boards from five to fifteen members. An unofficial mediator suggested cumulative voting, which black and white leaders agreed was fairer than limited voting, already in use in several Alabama cities. The compromise called for boards of seven members running at large. An odd number was needed to avoid tie votes, and seven seemed about right. Although the plan did not guarantee African Americans a seat, they could elect a board member by voting as a bloc for a single candidate sufficiently popular to attract a few white voters.[27] Skeptics grumbled, but the court approved the settlement.

Residents quickly learned how plumping works. In the first election under cumulative voting, most black voters cast all seven votes for a black candidate, Bobby Agee, who bagged a seat on the county commission. Four years later Agee won reelection despite the candidacy of a second black, who might have split the African American vote. Legal scholars Richard Pildes and Kristen Donoghue, who studied Chilton County's first seven years with alternative voting, found many benefits and few problems.[28] Minority residents they spoke with identified three important gains: improved road paving in black neighborhoods, the unprecedented appointment of African Americans to the Hospital Board and the Water Board, and an enhanced pride and sense of participation. Underscoring Lani Guinier's point that "cumulative voting is race neutral," the new system also helped Republican and female candidates, who had fared poorly in winner-take-all elections.[29] As expected, more candidates ran and more voters turned out. Despite an electorate strongly polarized along racial lines, the county's experience with cumulative voting did not confirm critics' fears of extremist victors, geographically concentrated winners, and more costly campaigns.

Alamogordo, New Mexico, also demonstrates that voters adapt quickly.[30] In 1987, local officials settled a voting-rights lawsuit by adopting cumulative voting for the three at-large positions on their seven-member city council. The four remaining members represent single-member districts, including a minority-majority district represented by a black councillor. Hispanics, who composed 24 percent of the population, sought representation on the council. No Hispanic had won an at-large seat since 1968, and Hispanics were too geographically dispersed to form a modestly compact minority-majority district. Community leaders agreed to try cumulative voting for the three at-large positions—a risky compromise if group members failed to plump their votes on a single Hispanic candidate or if their candidate of choice failed to attract more than a handful of white voters. Their faith was warranted, thanks in part to an intense voter-education campaign: a Latina won a seat on the council, and an exit poll revealed that 64 percent of Hispanics cast all three votes for a single candidate, in contrast to only 40 percent of Anglos and blacks.[31]

A study of sixteen small Texas municipalities confirmed the importance of voter education and white crossover.[32] As in Alamogordo, Anglo politicians eager to settle a voting-rights lawsuit accepted cumulative voting as an alternative to single-member districts. Exit polling indicated that nine out of ten black and Hispanic voters understood plumping and preferred the new system. By contrast, two in ten Anglos considered cumulative voting confusing, and a quarter believed it grossly unfair. Although bloc voting was rampant, minority candidates generally were successful where they did not have to depend greatly, if at all, on white voters. But in seven of the sixteen jurisdictions, no minority candidate won: in some cases multiple candidates split the minority vote, and in others the minority population was simply too small to elect one of its own. Small minority populations are at a distinct disadvantage in localities that elect half the city council or school board in alternate elections.

Despite these drawbacks, cumulative voting addresses Lani Guinier's key point: "51 percent of the people should not always get 100 percent of the power; 51 percent of the people should certainly not get all the power if they use that power to exclude the 49 percent. In that case we do not have majority rule. We have majority tyranny."[33] And as Guinier notes, cumulative voting promotes cross-racial coalitions and obviates gerrymandering.

Cumulative voting might resolve vote-dilution disputes, but can the courts impose it unilaterally? Perhaps not, as federal district judge Joseph Young discovered in 1995. The previous year he had found that at-large voting in Worcester County, Maryland, unlawfully diluted the votes of African Americans, who composed 21 percent of the county's population but lacked a seat on the five-member Board of Commissioners.[34] County voters had never elected a black official in 253 years, and the sitting commissioners, who eventually spent over $1 million fighting the lawsuit, were not eager to make amends.[35] Both sides presented redistricting plans, but none was suitable. White bloc voting undermined the minority-influence district suggested by the county, and African Americans were too widely dispersed for a suitably compact minority-majority district. Also, most county residents apparently preferred at-large voting. In the wake of Supreme Court wariness about racial gerrymanders, cumulative voting seemed a reasonable compromise.[36] As the judge opined in his decision, "Cumulative voting, unlike single-member districts, will allow the voters, by the way they exercise their votes, to 'district' themselves based on what they think rather than where they live."[37] County officials appealed, and the appellate court ruled that the judge had abused his authority. So the case went back to the district court, where Judge Young ordered elections based on five single-member districts, one with a 58 percent black majority.[38] As figure 10.1 reveals, the alternative to cu-

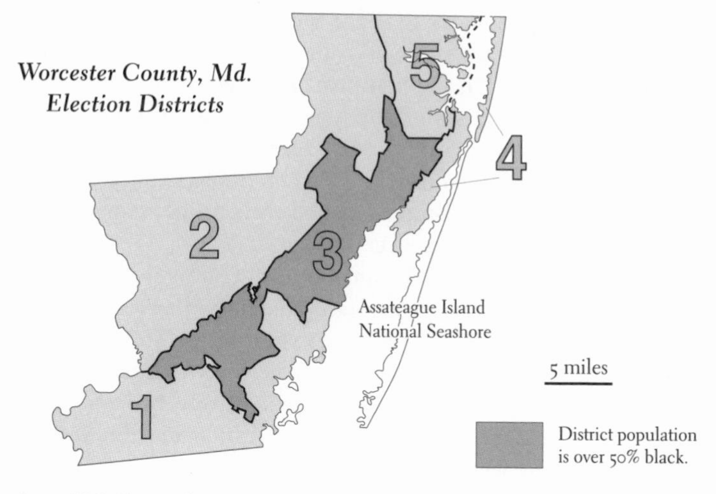

Figure 10.1. Election districts, Worcester County, Maryland. Compiled from map distributed by the Worcester County Board of Elections.

mulative voting was a complex district linking minority neighborhoods in three communities while skirting largely white coastal developments in the southeast. It worked, though, and the county elected its first black commissioner.

What are the risks of letting voters "district themselves"? Those who don't share Judge Young's perspective raise fears of racially polarized voting. Shortly after his decision, the *Baltimore Sun* denounced his decision because cumulative voting "will encourage all blacks in the county to vote only for a black candidate, and all whites to vote only for white ones. This will lead black campaigners to ignore white voters and white campaigners to ignore black voters. At least in a majority-black district, black candidates would have to compete for white votes, and in majority-white districts, white candidates would have to seek black votes."[39] The judge's critics were not as bitter as Lani Guinier's. *Newsweek* columnist George Will labeled her ideas "extreme, undemocratic and anticonstitutional,"[40] and Utah's conservative senator Orrin Hatch condemned Clinton's nominee as the "architect of a theory of racial preferences that if enacted would push America down the road of racial balkanization."[41] Even the *New Republic,* an advocate of cumulative voting, labeled her "a firm believer in the racial analysis of an irreducible, racial 'us' and 'them' in American society."[42]

But it's not just a matter of race. Critics caution that cumulative voting and other forms of proportional representation can empower extremist groups from all parts of the political spectrum, including anti-Semites and Uzi-carrying survivalists as well as welfare rights proponents and Black Power separatists. In an editorial titled "Proportional Representation Fails," the *New York Times* warned of corrupt negotiation among a multitude of splinter parties like those in Italy and Poland and chided Israel for having "replicated the proportional system used by the Weimar Republic, whose disastrous failure opened the way to Hitler's takeover."[43]

Are these fears warranted? My map (fig. 10.2) of countries with a form of proportional or preference voting suggests otherwise. Democratic alternatives to plurality or majority voting are simply too widespread to dismiss the method as incomprehensibly arcane or inherently fractious. Proportional voting is used extensively throughout the world, by developed countries in northern Europe and the western Pacific as well as by less prosperous nations in Latin America and parts of Africa. By contrast, a second map (fig. 10.3) shows countries with plurality or majority voting concentrated in North America,

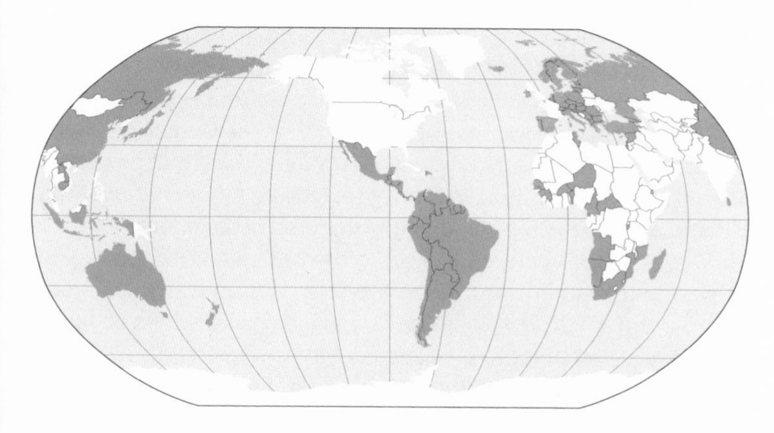

Figure 10.2. Countries with a form of proportional representation, including mixed, semiproportional, and preference voting. Generalized from the Electoral Systems Survey compiled by Karen Taggart for the Center for Voting and Democracy.

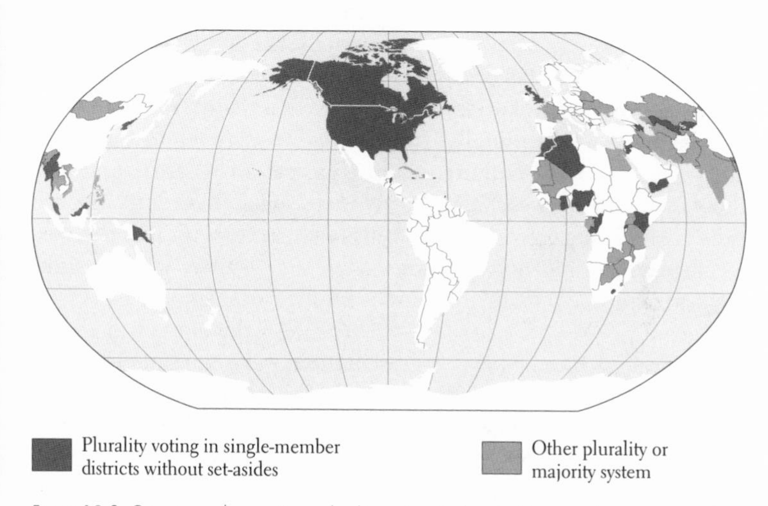

Figure 10.3. Countries with majority or plurality voting. Darker shading identifies countries with single-member districts and plurality voting without set-asides. Generalized from the Electoral Systems Survey compiled by Karen Taggart for the Center for Voting and Democracy.

Africa, and southwestern Asia. The only nations in Western Europe with a voting method similar to the American way are France and Britain. Unfortunately, nearly a fifth of the nations listed are too tiny for a pair of very small scale world maps. My first map thus ignores the likes of Andorra and Cyprus, while its companion omits the Seychelles, much of the Caribbean, and other small nations.

The maps, I must concede, oversimplify the comprehensive list of 191 countries compiled by Karen Taggart at the Center for Voting and Democracy.[44] A concise characterization of electoral procedures around the world, Taggart's catalog included a number of countries with no elections as well as several with exotic descriptions like "people's congresses" (which Taggart placed in quotation marks) for Libya, "in transition" for Eritrea, and "??" for Guinea-Bissau, North Cyprus, and Togo. Despite a dozen or so anomalies, most countries have an electoral process, even such dubiously democratic nations as North Korea, which has plurality elections in single-member districts—the system used in the United States.

The maps also obscure important distinctions like the "set-asides" used to reserve seats for ethnic minorities in Botswana, India, and nineteen other countries with majority or plurality voting. The most common electoral system represented on figure 10.2 is the "party list" system of proportional representation, based on multimember constituencies, but the same gray shading covers countries with mixed, semiproportional, and preference voting systems. Although figure 10.3 is a similar stew of electoral complexity, the darker gray highlighting nations with plurality voting in single-member districts without set-asides affords a more accurate portrait of where people vote the way we do. However flawed, my maps make two points: American-style elections are not a prerequisite for democracy, and the various forms of proportional representation are far more widespread—and ostensibly effective—than critics imply.

How would proportional representation work in American congressional elections? Quite well, I think, judging from the map PR activist Lee Mortimer designed to help North Carolina address minority voting rights without bizarre racial gerrymanders.[45] As figure 10.4 illustrates, comparatively straightforward boundaries partition the state into three multimember districts without splitting any counties. The boundaries reflect the state's physical and historical geography, which underlies economic and social issues important to district residents. For example, the Eastern District, largely on the Coastal Plain's sandy soils and flattish terrain, is more concerned with agriculture and coastal flooding than the Piedmont District, in the foothills of the Appalachians, where hilly terrain and water power left a legacy of small farms and manufacturing. By contrast, residents of

Multimember districts for cumulative voting

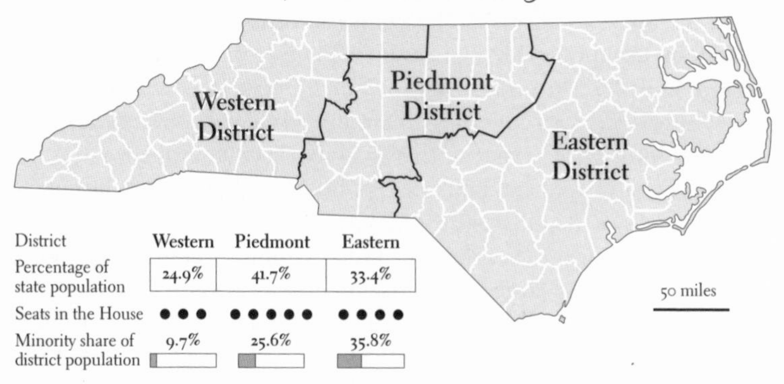

Figure 10.4. Multimember districts designed by Lee Mortimer for cumulative voting in North Carolina. Map redrawn from illustration in the *Congressional Quarterly Weekly Report* 52 (1994): 813.

the Western District, in the Blue Ridge Mountains, a region too rugged and remote for plantation agriculture, have a greater stake in Appalachian redevelopment and the management of federal forest lands. With the smallest population, the Western District would elect only three of the state's twelve House members, whereas the Eastern and Piedmont Districts would control four and five seats, respectively. In addition to geographically significant boundaries, the plan has a low disparity ratio (1.006), with population per seat varying inconsequentially, from 551,000 to 554,000 people.[46]

If linked to cumulative voting, the map could have sent African Americans Eva Clayton and Melvin Watt to Congress without infringing the constitutional rights of Robinson Everett and his fellow plaintiffs in *Shaw v. Reno.* As the graph in figure 10.4 indicates, North Carolina's largely black minority population is sufficiently numerous to control single seats in the Eastern and Piedmont Districts—without the need for filler people, thanks to plumping. Syndicated columnist William Raspberry, an influential black advocate, applauded "superdistricts" and cumulative voting as going "a long way toward promoting cooperation across party, racial and other lines, since candidates would have to appeal to a much wider electorate [and] demagogues would have a tough time winning."[47] Eager for a compromise, Raspberry gloated that both Watt and Everett endorsed the plan.

Mortimer's proposal never got the hearing it deserved. In late 1967 Congress had passed H.R. 2275, an amended immigration bill making Philippines native Dr. Ricardo Vallejo Samala a permanent resident and requiring single-member districts for all states with more than one House member.[48] The House and the Senate had debated a variety of electoral reforms in the wake of *Wesberry v. Sanders* but agreed on little. The end-of-session amendment to Dr. Samala's private bill was a desperate effort to circumvent court-ordered at-large elections in eighteen states that had not yet complied with the Supreme Court's one person, one vote doctrine. Better that the courts redraw the lines than disrupt the careers of half the House.[49] The bill also addressed fears that southern states might defang the Voting Rights Act with at-large, winner-take-all congressional elections.[50] Proportional representation was little more than a failed effort to reform municipal elections, and apparently no one considered that the bill might someday prove an obstacle to cumulative voting in multimember districts.

Congressional supporters of proportional representation tried persistently to remove the roadblock. In 1995 Georgia congresswoman Cynthia McKinney, threatened with the loss of her district by the high court's decision in *Miller v. Johnson,* introduced the Voters' Choice Act.[51] McKinney tried again in 1997. A year later, Illinois Citizens for Proportional Representation sought (unsuccessfully) a sponsor for a similar bill named the Illinois Option Act, to commemorate 110 years (1870-1980) of cumulative voting in multimember districts of the lower house of the Illinois legislature.[52] And in 1999 North Carolina congressman Melvin Watt introduced the States' Choice of Voting Systems Act.[53] All these bills were clear and concise, all would let states set up multimember districts, and all failed.

It's a cause worth pursuing. As political essayist Michael Lind pointed out in the *Atlantic Monthly* nine months before Lani Guinier's abortive nomination, proportional representation is more democratic than other proposed reforms, including term limits, a stronger presidency, and a British-style parliament.[54] It's not just about race, after all, and it's not the dire threat its opponents contend. As former presidential candidate John Anderson and PR activist Robert Richie argue, "for every voter who wants to elect an extremist, there are many more voters denied an opportunity to elect a more centrist representative."[55] It would work as well as bushmanders in empowering racial minorities and Republicans.

If Americans ever adopt proportional representation (ideally preference voting, as in Cambridge), I hope we take the next logical step and embrace weighted voting in the House of Representatives. More radical than PR perhaps, weighted voting satisfies the one person, one vote requirement by adjusting a district's electoral clout to its population.[56] For example, if the population of a three-member district warrants only 2.7 seats, each representative will have only 0.9 votes. And if a district is entitled to 3.3 seats, its three representatives, like shareholders in a corporation, will each cast 1.1 votes. If this seems unfair or unworkable, consider the traditional legislative committee structure, which accords committee chairs and senior members of the dominant party greater leverage than their junior colleagues and far more weight than members of the minority party.[57] As long as the weights do not vary radically—a range between 0.65 and 1.3 seems workable—the system is equitable and understandable.

Weighted voting has two advantages over the equal-representation myth: temporal stability and regional integrity. Because the weights compensate for differences in district population, geographically significant boundaries need not change every ten years. Equally important, redistricting committees need not assign counties to less appropriate districts just to keep the disparity ratio low. If proportional representation is worth adopting, it's worth protecting from arbitrary boundaries drawn to equalize populations while simultaneously appeasing incumbents, party bosses, and protected minorities.

11 *Epilogue*

OUR EXAMINATION OF THE AMERICAN way of redistricting raises three issues: Should race matter, should shape matter, and should geography matter. The post-2000 remap and the prospect of ever more outrageous congressional districts—but with less overt concern for ethnicity—raises a fourth, perhaps more fundamental question: What next?

The answer to the race question is so obviously affirmative (Who can deny the salient divisiveness of race in American society?) that the query needs rephrasing, with emphasis on *how* rather than *whether.*[1] Indeed, since the early 1950s Congress and the courts have repeatedly wrestled with *how* race should matter. In the process we buried Jim Crow, as a century earlier we dismantled slavery, and during the 1990s we demonstrated that a remap designed to encourage minority candidates need not injure white voters. Given most incumbents' irresistible urge to curry favor with all constituents, regardless of color, was anyone truly surprised when political scientist David Canon found that black-majority districts work quite well in representing whites?[2] But in increasing the electability of minority candidates we have produced weird silhouettes that invoke public ridicule and judicial reprimand—shapes like the North Carolina district that white plaintiffs denounced as "harmful" and Justice Sandra Day O'Connor labeled "bizarre."

If minority-majority districts are at all harmful, the more likely victims are African Americans and Hispanics. That's the verdict of social scientists who question the effects of minority-majority districts on the "substantive representation" of minority groups, which refers to the groups' clout in getting laws passed and funding approved.[3] Using the remap to increase "descriptive representation"—getting more people of color into House seats—concentrates minority voters into comparatively few black-majority or Hispanic-majority districts, thereby undermining minority support for white Democrats, their traditional allies. And a Congress with an increased number of minority representatives might well be a Congress dominated by Republicans, who are less likely to promote policies favored by minority-group leaders. At least that's the apparent result of the Republican takeover of the House in the 1994 elections and Congress's subsequent retreat from affirmative action into welfare reform and other tenants of the GOP's Contract with America. With time, though, stronger black and Hispanic caucuses and more experienced minority incumbents might prove a worthwhile investment.

Should shape matter? Not really, argue legal scholars like Pamela Karlan, who defends bizarre districts by noting the courts' inability "to identify a concrete harm to any identifiable individual."[4] In a similar vein, political scientist Micah Altman, who examined the effects of geometrically irregular minority-majority districts on public confidence, found "no support for the hypothesis that 'ugly' districts send pernicious messages to voters that affect their attitudes toward government or Congress."[5] Quoting a phrase rampant in recent redistricting lawsuits, Altman notes, "The only detectable effect of shape was on turnout. Moreover, I could find no evidence that bizarre districts cause 'expressive harms.'"[6] And who can argue that the weirdly shaped or questionably contiguous districts drawn to promote minority representation are more difficult to represent than the far larger districts stretched across sparsely populated reaches of the American West? After all, dispersed minorities in parts of Chicago, New York City, or the rural South share common concerns much like the coastal issues that unite island residents with their mainland neighbors.[7] Both kinds of districts constitute communities of interest in which compactness and contiguity have little relevance.

Shape will continue to matter, though, because highly irregular districts, at odds with how most Americans think elections districts should look, are a form of cartographic mischief. For this reason, tiny silhouette maps, with little relevance to how well a district can be represented, are enormously effective as propaganda against graphically flagrant gerrymanders of any sort, partisan as well as racial. And when the judiciary targets only racial gerrymanders, as in the 1990s, Rorschach-like silhouettes even provide a clever Why me? defense of deposed bushmanders: one of the most effective arguments supporting the Texas minority-majority districts struck down in *Bush v. Vera* was a poster that juxtaposed four anonymous silhouettes and dared the viewer to separate two districts overturned by the Court from two equally irregular white-majority districts allowed to stand. Most viewers, I suspect, wondered why the Supreme Court didn't reject them all.

Especially irksome are districts that challenge the contiguity convention with either quasi outliners connected only at a point (like the portion of North Carolina's Twelfth District in fig. 3.5) or similarly suspicious inclusions from a neighboring district (like the part of District 2 in the same illustration).[8] Although judges tolerate the questionable contours of point contiguity, redistricting officials cautiously avoid distinctly disconnected outliers, except where islands or other coastal configurations offer a plausible excuse. So strong is the outlier taboo that figure 11.1 will surely surprise if not shock most jurists and legislators by revealing New York's persistent delineation of a noncontiguous district in the upper Hudson–Lake Champlain area in the early decades of the republic. But as geographer Kenneth Martis confirms in his exhaustive compilation of congressional district boundaries, flagrantly fragmented districts were clearly an option until Congress introduced the contiguity and compactness standards in 1842.[9] If redistricting committees choose to delineate communities of interest without the subterfuge of tentacle-like appendages—and if the courts choose to let them—there is ample historical precedent to counter critics.

If the Constitution mandates neither compactness nor contiguity,

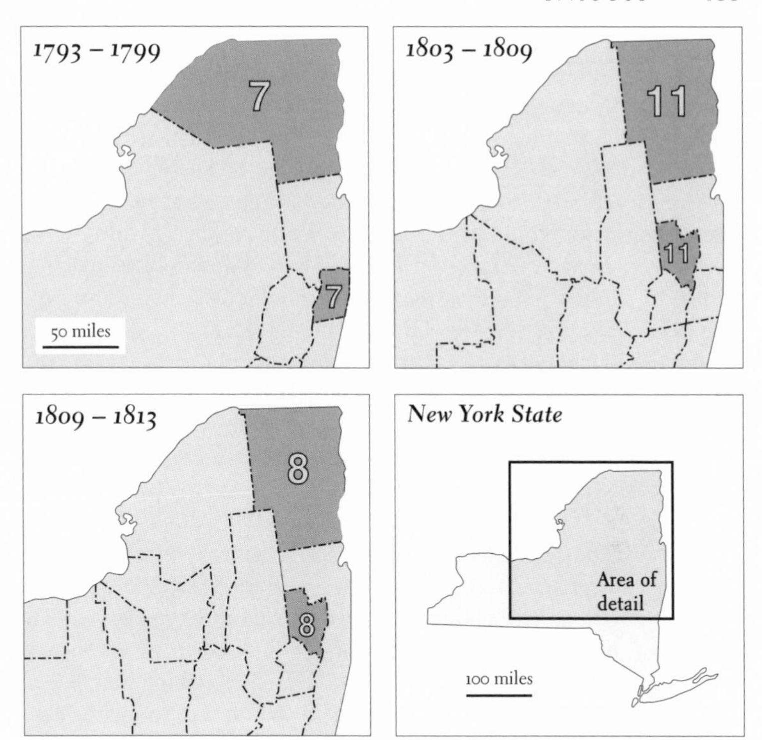

Figure 11.1. Before Congress enacted a contiguity standard in 1842, the New York legislature repeatedly established a noncontiguous congressional district in the northeastern part of the state. Lines are congressional district boundaries. Compiled from Kenneth C. Martis, *The Historical Atlas of United States Congressional Districts, 1789-1983* (New York: Free Press, 1982).

should geography matter? Absolutely, as long as we understand that geography means more than a simplistic sense of shape and distance. Because of vastly improved transport and communication, the geography that's relevant for the twenty-first century is a very different from the geography of 1790, 1920, or even 1960. And though the slogan "all politics is local" remains valid, proximity plays a very different role today than in the heyday of ward politics and strong party loyalty.[10] Demographic affinity—nothing new, really—is more germane to most concerns than neighborhoods are, and even neighborhood issues like crime and zoning frequently precipitate local alliances among geographically dispersed groups like elderly residents and upper-middle-class homeowners. Although African Americans, Hispanics, born-again Christians, young urban professionals, and other demographic clusters might have much in common with their neighbors, communities of interest are almost always larger and more fragmented than one's immediate neighborhood.[11] As a result, traditional district boundaries, whether for congressional, state legislative, or city council districts, do not work as well as they once did.

Can the contorted boundaries drafted in the early 1990s to equalize district populations and protect coherent minority voting rights really work better than more traditional borders? Maybe, maybe not. But they're clearly less dysfunctional than silhouette maps and caustic critics suggest. North Carolina's I-85 District, for instance, strongly reflects in both its name and its shape the radical reduction of functional distance between its seemingly dispersed nodes. And intricate inner city districts like New York's Bullwinkle District or Chicago's "pair of earmuffs" are comparatively compact in a social-geographic space based on personal interaction rather than surveyor's instruments.

Even so, bushmanders typically lack what redistricting scholar Bernard Grofman calls "cognizability," defined as "the ability to characterize the district boundaries in a manner that can be readily be communicated to ordinary citizens of the district in commonsense terms based on geographical referents."[12] In denouncing North Carolina's I-85 District, he condemned all "districts that run helter-skelter the course of a state, picking up noncontiguous pockets of minorities [and] cutting up cities as with a scapel . . . [in ways that] may even . . . violate due process or equal protection."[13] Careful to differentiate "non-cognizability" from "ill-compactness," which is not necessarily objectionable, Grofman listed several infringements of the "cognizability principle," including "violation of natural geographic boundaries, grossly unnecessary splitting of local subunit boundaries (such as city and county lines), and sunderings of proximate and natural communities of interest."[14] In short, citizens should not have to consult a 1,272-page atlas to find out who represents them.

Cognizability has three natural enemies: artificial enhancement of minority representation, exact population equality, and unbridled partisan greed. Although judges routinely rejected Justice Department efforts to maximize the number of minority-majority districts at the expense of compactness, the courts actively support balanced populations and rarely interfere with the leading party's cartographic prerogatives. In the contentious saga of North Carolina's I-85 District, described in chapter 3, the Supreme Court underscored this point in a unanimous 1999 decision reversing a lower court that struck down the state's 1997 remedial remap because of racial motives. The earlier ruling was flawed, the justices ruled, because partisan objectives eclipsed racial results.

Another enemy of cognizability is the ease with which political cartographers can exploit block data and geographic information systems. As University of Texas law professor Samuel Issacharoff notes, "with the advent of the computer, sophisticated political actors showed remarkable resourcefulness in gerrymandering political boundaries without running afoul of the equal population principle."[15] Single-member districts are especially vulnerable to violations of cognizability. "Unless courts move in the direction of specific compactness constraints," Issacharoff argues, "the search for reasonable configurations of districts will become a distinct liability for territorially based electoral systems."[16]

Multimember districts with proportional representation are an obvious remedy. As discussed in chapter 10, any one of several proportional representation plans could simultaneously put to rest concerns about cognizability, minority voting rights, and population equality. If minority voters or members of any other group want to plump their ballots on a narrowly focused candidate, fine—let them. It's better to give African Americans, Hispanics, and even white racists a choice rather than corral them into contorted districts certain in any event to exclude at least a few group members. And if minority voters choose not to back a group candidate, that's fine too. As for critics' fears of divisively fragmented legislatures, it's hardly likely that proportional representation will inundate Congress with platoons of Al Sharptons and David Dukes. I'd be surprised, though, if regions with minority-majority districts in the post-1990 remap (fig. 11.2) didn't elect roughly similar numbers of minority House members. Multimember districts that allow a voluntary expression of ethnic geography are arguably fairer than the highly intricate boundaries required for single-member districts.

Weighted voting could confer a further benefit: geographic stability. As described in chapter 10, weighted voting by elected representatives can honor the one person, one vote requirement and allow otherwise objectionable variations in district population. With this flexibility, districts could reflect natural, inherently stable geographic boundaries, and the decennial census might be used merely to readjust weights, not reconfigure boundaries. It's a better way, but don't hold your breath.[17]

What's certain is a shift of approximately eleven House seats after

Black-majority districts

Hispanic-majority districts

Figure 11.2. Minority-majority districts in the 103rd Congress, 1993-94. Compiled from list in *Congressional Quarterly Weekly Report* 51 (1993): 1763.

the 2000 census. Although "equal proportions" reapportionment might well produce a few cliff-hangers, as it did in 1990, demographers expect Arizona, Georgia, and Texas to gain two seats each, while redistricting officials in New York and Pennsylvania face a remap with two fewer districts. Completing the exchange, Colorado, Florida, Montana, Nevada, Utah, and perhaps California should each pick up a seat, while Connecticut, Illinois, Michigan, Mississippi, Ohio, Oklahoma, and Wisconsin will each redistrict with one representative fewer.[18]

What's not clear is whether redistricting committees will push partisan gerrymandering well beyond the public's begrudging tolerance of 1990s-style bushmanders and Bullwinkles. Although GIS technology stands ready to craft some truly outrageous silhouettes, contiguity measures embedded in the software can help lawmakers and judges resist the trend. If political cartographers don't exercise restraint, openly scandalous assaults on cogizability might eventually move the nation toward multimember districts and proportional representation. Then again, awareness of the potential abuses of GIS and the power of cartographic propaganda might move state legislatures and redistricting officials to a more conservative process, if not a more open and inclusive one, guided by public participation as well as a compactness constraint, formal or informal. In addition to demonstrating that the technology is indeed neutral, this latter strategy could reinvigorate the nation's much-abused system of single-member districts.

NOTES

CHAPTER ONE

1. James T. Austin, *The Life of Elbridge Gerry, with Contemporary Letters to the Close of the American Revolution* (Boston: Wells and Lilly, 1829; reprint, New York: Da Capo Press, 1970), 2:347.

2. Accounts of the origin of the gerrymander attribute the name to various editors or politicians and the image to different artists, among whom Elkanah Tisdale is the most likely originator. See John Ward Dean, "The Gerrymander," *New England Historical and Genealogical Register* 46 (1892): 374–83.

3. See, for example, Jonathan P. Hicks, "Albany Lawmakers Agree on Redrawn 12th District," *New York Times*, July 31, 1997, B3.

4. Dave Kaplan, "Constitutional Doubt Is Cast on Bizarre-Shaped District," *Congressional Quarterly* 51 (1993): 1761–63, quotation on 1761; Paul V. Niemeyer, "The Gerrymander: A Journalistic Catch-Word or Con-

stitutional Principle? The Case in Maryland," *Maryland Law Review* 54 (1995): 242-60, quotation on 246; and Adam Clymer, "Republicans Gain as States Battle over Redistricting," *New York Times,* December 30, 1991, A1.

5. Bernard Grofman, Lisa Handley, and Richard G. Niemi, *Minority Representation and the Quest for Voting Equality* (New York: Cambridge University Press, 1992), 16-19.

6. For interpretations of how the Voting Rights Act prohibits vote dilution, see ibid., 29-31; Bernard Grofman, Robert Griffen, and Amihai Glazer, "The Effect of Black Population on Electing Democrats and Liberals to the House of Representatives," *Legislative Studies Quarterly* 18 (1992): 365-79; and David Lublin, *The Paradox of Representation* (Princeton: Princeton University Press, 1997), 6-9, 120.

7. Charles S. Bullock III, "Winners and Losers in the Latest Round of Redistricting," *Emory Law Journal* 44 (1995): 943-77; Lisa A. Kelly, "Race and Place: Geographic and Transcendent Community in the Post-*Shaw* Era," *Vanderbilt Law Review* 49 (1996): 227-308; Charles Mahtesian, "North Carolina Tries Again; Georgia Map Rejected," *Congressional Quarterly* 50 (1992): 188-89; and Ripley Eagles Rand, "The Fancied Line: *Shaw v. Reno* and the Chimerical Racial Gerrymander," *North Carolina Law Review* 72 (1994): 725-58.

8. See Kevin Sack, "Redistricting Plans Approved; 2 Rulings Clear Revisions for Albany and Congress," *New York Times,* July 3, 1992, B1. Because out-migration cost New York three House seats, the state's post-1990 reapportionment was far from harmonious. For a concise account of the partisan bickering and court challenges that preceded submission of a acceptable plan, see Mark Monmonier, *Drawing the Line: Tales of Maps and Cartocontroversy* (New York: Henry Holt, 1995), 209-12.

9. Joe Fodor, "Kings County Almanac: Reapportion This!" *Brooklyn Bridge* 2 (July 1997): 20-21.

10. For the full text of Chapter 137, which describes the state's new congressional districts, see New York Consolidated Laws Service, *1992 Session Laws: Chapters 1-537* (Rochester, N.Y.: Lawyers Cooperative Publishing, 1993); quotation on 1181, as amended by the "technical correction," published as Chapter 138, which (on 1234) changed "New York avenue" in the original version to "Jamaica avenue."

11. U.S. Bureau of the Census, *Congressional District Atlas: 103rd Congress of the United States* (Washington, D.C., 1993).

12. Lubin, *Paradox of Representation,* 47-48, 99. In response to federal pressure, Mississippi redrew its congressional districts in 1984 to provide a district around Jackson with a black voting-age majority. Even so, the district did not elect an African American until 1986, when a vigorous get-out-the-vote effort helped send Mike Espy to Washington. The post-1990 remap increased the district's black majority from 58 to 63 percent and helped ensure the election of a black successor, in a special election held in 1993 after President Clinton appointed Espy secretary of agriculture.

13. Dean, "Gerrymander."

14. Some political scientists question whether (or to what extent) minority-majority districts actually helped Republican candidates. See, for example, Kimball Brace, Bernard Grofman, and Lisa Handley, "Does Redistricting Aimed to Help Blacks Necessarily Help Republicans?" *Journal of Politics* 49 (1987): 169-85.

15. John H. Fund, "Beware the Gerrymander, My Son," *National Review* 41 (April 7, 1989): 34-36; and Robert Benenson and others, *Jigsaw Politics: Shaping the House after the 1990 Census* (Washington, D.C.: Congressional Quarterly, 1990), 93-96.

16. A recent study of the shapes of congressional districts listed District 36 among twenty-eight districts with "low dispersion or perimeter compactness scores." But the district was among the least distorted in the table, and the measures seem to have been influenced by a coastal boundary that includes two offshore islands. See Richard H. Pildes and Richard G. Niemi, "Expressive Harms, 'Bizarre Districts,' and Voting Rights: Evaluating Election-District Appearances after *Shaw v. Reno*," *Michigan Law Review* 92 (1993): 483-587; table on 565.

CHAPTER TWO

1. House Committee on the Judiciary, *Voting Rights: Hearings before the Subcommittee on Civil and Constitutional Rights*, 103rd Cong., 1st and 2nd sess., 1993-94, 86.

2. Laurence F. Schmeckebier, *Congressional Apportionment* (Washington, D.C.: Brookings Institution, 1941).

3. Ibid., 74, 113. In 1842, Congress also reset the apportionment ratio to 76,680.

4. The two tables in this example are adapted from Schmeckebier, *Congressional Apportionment*, 75.

5. Ibid., 7; and Michel L. Balinski and H. Peyton Young, *Fair Representation: Meeting the Ideal of One Man, One Vote* (New Haven: Yale University Press, 1982), 37-43.

6. Balinski and Young, *Fair Representation*, 40-44. The committee also discovered another example of the Alabama paradox: the awkward instability in the size of Maine's delegation, which stood at three for a House with 350 to 382 seats, rose to four for 383 to 385 seats, dropped to three for 386 seats, jumped back to four for 387 or 388 seats, and fell back to three for 389 and 390 seats, before stabilizing at four for 391 to 400 seats.

7. Schmeckebier, *Congressional Apportionment*, 13-17. Wilcox also distinguished himself as a demographer, an economist, and a statistician.

8. Ibid., 120-24; and Charles W. Eagles, *Democracy Delayed: Congressional Reapportionment and Urban-Rural Conflict in the 1920s* (Athens: University of Georgia Press, 1990), 32-84.

9. Balinski and Young, *Fair Representation*, 46-59; and Dudley L. Poston Jr., "The U.S. Census and Congressional Apportionment," *Society* 34 (March–April 1997): 36-44.

10. Schmeckebier, *Congressional Apportionment*, 133.

11. Ibid., 134. Quotation is from the report of a select committee appointed to

study Adams's resolution attacking Tyler. Adams was a member of the committee and an influential author of its report, which supported the resolution.

12. An Act for the Apportionment of Representatives to Congress among the Several States according to the Ninth Census, *U.S. Statutes at Large* 17 (1872): 28.

13. An Act Making an Apportionment of Representatives in Congress among the Several States under the Twelfth Census, *U.S. Statutes at Large* 31 (1872): 733, 734.

14. According to Charles Eagles, embarrassment moved Congress to pass the 1929 apportionment act. See Eagles, *Democracy Delayed,* 83-84.

15. *Wood v. Broom,* 287 U.S. 1, 8 (1932).

16. Broom's apparent goal was to run at large in a statewide election, a distinct possibility if the Court voided the new boundaries a few months before the election.

17. *Broom v. Wood,* 1 F. Supp. 134, 135 (S.D. Miss. 1932).

18. *Broom v. Wood,* 138.

19. Taking a states' rights position, Mississippi's lawyers ignored the pivotal argument, advanced in an amici curiae brief filed by Virginia's attorney general and his associates, that the 1911 statute no longer held. See *Wood v. Broom,* 77 L. ed. U.S. 131, 133 (1932).

20. *Broom v. Wood,* 137.

21. Stanley B. Parsons, Michael J. Dubin, and Karen Toombs Parsons, *United States Congressional Districts, 1883-1913* (New York: Greenwood Press, 1990), 366-67.

22. Although the number of Americans living in rural areas increased from 46 to 62 million between 1900 and 1950, many counties, especially in the South and Midwest, registered their maximum population in 1900 or earlier. By contrast, over this fifty-year period the number of urban residents jumped from 30 to 89 million. The population of cities of 250,000 or more grew from 11 to 35 million residents, and the number of cities with populations greater than 500,000 rose from six to nineteen. Between 1920 and 1950, when the size of the average American household dropped from 4.34 to 3.52 persons, the number of farm households declined from 6.8 to 6.3 million. See U.S. Bureau of the Census and Social Science Research Council, *The Statistical History of the United States from Colonial Times to the Present* (Stamford, Conn.: Fairfield, 1965), 14-16; and U.S. Bureau of the Census, *1970 Population as a Percent of Maximum Population by Counties of the United States,* GE-50 series, map 43 (Washington, D.C., 1972).

23. Andrew Hacker, *Congressional Districting: The Issue of Equal Representation* (Washington, D.C.: Brookings Institution, 1963), 94.

24. Ibid., 23.

25. Ibid., 106-10.

26. *Baker v. Carr,* 369 U.S. 186 (1962). Also see Charles S. Rhyne, "The Progeny of Baker v. Carr," in *Legislative Apportionment: Key to Power,* ed. Howard D. Hamilton (New York: Harper and Row, 1964), 65-70.

27. U.S. Constitution, amend. 14, sec. 1.

28. *Baker* in effect reversed an earlier decision, made in 1946, when the Supreme Court declared the federal courts unfit to judge the fairness of congressional redistricting. See *Colegrove v. Green,* 328 U.S. 549 (1946).

29. *Wesberry v. Sanders,* 376 U.S. 1, 18 (1964). District populations for the 1960 census ranged from 823,680 to 272,154.

30. For a review of the Supreme Court's decreased tolerance of deviations from the equal-population standard, see Robert G. Dixon Jr., "The Warren Court Crusade for the Holy Grail of 'One Man–One Vote,'" in *Supreme Court Review, 1969,* ed. Philip B. Kurland (Chicago: University of Chicago Press, 1969), 219-70.

31. *Wells v. Rockefeller,* 394 U.S. 542 (1969). District populations for the 1960 census ranged from 435,880 to 382,277.

32. *White v. Weiser,* 412 U.S. 783 (1973). District populations for the 1970 census ranged from 477,856 to 458,581.

33. *Mahan v. Howell,* 410 U.S. 315, 319 (1973).

34. *Karcher v. Daggett,* 462 U.S. 725, 744 (1983). District populations for the 1980 census ranged from 527,472 to 523,798.

35. U.S. Constitution, amend. 15.

36. See David T. Canon, *Race, Redistricting, and Representation: The Unintended Consequences of Black Majority Districts* (Chicago: University of Chicago Press, 1999), 63-70; Chandler Davidson, "The Voting Rights Act: A Brief History," in *Controversies in Minority Voting: The Voting Rights Act in Perspective,* ed. Bernard Grofman and Chandler Davidson (Washington, D.C.: Brookings Institution, 1992), 7-51. For a comparatively conservative viewpoint, see Stephen Thernstrom and Abigail Thernstrom, *America in Black and White: One Nation, Indivisible* (New York: Simon and Schuster, 1997), 462-92.

37. Frank R. Parker, *Black Votes Count: Political Empowerment in Mississippi after 1965* (Chapel Hill: University of North Carolina Press, 1990), 31, 200.

38. John D. Feerick and others, *Congressional Redistricting* (Washington, D.C.: American Bar Association, Division of Public Service Activities, 1981), 28; and 2 U.S.C. § 2c (1967).

39. Laughlin McDonald, "The 1982 Amendments of Section 2 and Minority Representation," in *Controversies in Minority Voting,* ed. Bernard Grofman and Chandler Davidson (Washington, D.C.: Brookings Institution, 1992), 66-84.

40. Armand Derfner, "Vote Dilution and the Voting Rights Act Amendments of 1982," in *Minority Vote Dilution,* ed. Chandler Davidson (Washington, D.C.: Howard University Press, 1984), 145-63.

41. *Thornburg v. Gingles,* 478 U.S. 30 (1986).

42. *Thornburg v. Gingles,* 50-51. Specifically, the high court established a three-pronged test for vote dilution in multimember districts: "First, the minority group must be . . . sufficiently large and geographically compact to constitute a majority of a single-member district . . . .Second, the minority group must be . . . politically cohesive . . . .Third, the white majority [must vote] as a bloc to . . . defeat the minority's preferred candidate." For a review of complications arising

from the *Gingles* decision, see Bernard Grofman, Lisa Handley, and Richard G. Niemi, *Minority Representation and the Quest for Voting Equality* (Cambridge: Cambridge University Press, 1992), 61-81.

43. *Davis v. Bandemer,* 478 U.S. 109 (1986).

44. See Federal Writers' Project of the Works Progress Administration, *Mississippi: A Guide to the Magnolia State* (New York: Viking Press, 1938), 99. Sponsored by the Mississippi Advertising Commission, the guidebook boldly describes the "economic slavery [of] the tenant-credit system."

45. Except for two counties (Holmes added in 1902 and Warren deleted the same year), the Third District established on April 5, 1956 (and not altered until April 3, 1962), was identical with the Delta District established on March 9, 1882; see Kenneth C. Martis, *The Historical Atlas of United States Congressional Districts, 1789-1983* (New York: Free Press, 1982), 242-43. At least one source is slightly inaccurate in describing the district as "preserved intact from 1882 to 1962"; see Frank R. Parker, "Racial Gerrymandering and Legislative Reapportionment," in *Minority Vote Dilution,* ed. Chandler Davidson (Washington, D.C.: Howard University Press, 1984), 85-117, quotation on 90.

46. Martis, *Historical Atlas,* 243. The new map established on April 5, 1956, merely transferred Calhoun County from District 2 to District 1.

47. Parker, *Black Votes Count,* 41-43.

48. According to data for 1960, the largest and smallest districts before the merger had 460,100 and 237,887 inhabitants, respectively. After reapportionment, the combined district had a population of 608,441, whereas the smallest district now held 295,072 residents. U.S. Bureau of the Census, *Congressional District Data Book (Districts of the 87th Congress)—A Statistical Abstract Supplement* (Washington, D.C., 1961), 34.

49. The legislature knew it had to act fast. In October 1965, a year after *Wesberry v. Sanders,* in which the Supreme Court clarified the need for a one person, one vote standard for congressional districts, the Mississippi Freedom Democratic Party, a civil rights movement within the Democratic Party, challenged the Mississippi 1962 remap in federal court. With the court's intention abundantly clear, the legislature reconfigured the state's congressional districts before the lawsuit, *Conner v. Johnson,* could come to trial. See Parker, *Black Votes Count,* 38, 43, and 85-91.

50. Ratio computed from population data in ibid., 52. According to 1960 census counts, the populations of districts delineated in 1966 varied from 449,565 for District 5 to 423,300 for District 4.

51. Populations for the 1970 census varied from 451,981 for District 5 to 433,825 for District 1. U.S. Bureau of the Census, *Congressional District Data Book (Districts of the 93rd Congress)—A Statistical Abstract Supplement* (Washington, D.C., 1973), 272.

52. Parker, *Black Votes Count,* 40-41.

53. *Connor v. Johnson,* 386 U.S. 483 (1967).

54. Parker, *Black Votes Count,* 85-91, quotation on 89. Parker considered the

decision "a major default of the Warren Court, which was otherwise sensitive to civil rights issues" (90).

55. Ibid., 90.

56. Ibid., 91.

57. Ibid., 154-56.

58. *Kirksey v. Board of Supervisors of Hinds County, Mississippi,* 402 F. Supp. 658, 666 (S.D. Miss. 1975); and 528 F. 2d 536 (5th Cir. 1976). Parker describes several similarly outrageous uses of cracking, and two of stacking, in Frank R. Parker, "County Redistricting in Mississippi: Case Studies in Racial Gerrymandering," *Mississippi Law Journal* 44 (1973): 391-425.

59. *Kirksey v. Board of Supervisors of Hinds County, Mississippi,* 554 F. 2d 139, 151 (5th Cir. 1977).

60. *Kirksey v. Board of Supervisors,* 151. Later that year the Supreme Court affirmed the circuit court by declining to hear an appeal; see 434 U.S. 968 (1977).

61. Voting Rights Acts Amendments of 1982, *U.S. Statutes at Large* 96 (1982): 131, 134. When signed by President Reagan on June 29, 1982, the bill became Public Law 97-205. See "Voting Rights Act Extended, Strengthened," *CQ Almanac* 38 (1982): 373-77; and Frank R. Parker, "Changing Standards in Voting Rights Law," in *Redistricting in the 1990s: A Guide for Minority Groups,* ed. William P. O'Hare (Washington, D.C.: Population Reference Bureau, 1989), 55-65.

62. Although most states requiring preclearance work with the Department of Justice, the Federal District Court for the District of Columbia can also grant preclearance. In the post-1990 remap, for instance, the District of Columbia court, not the Justice Department, approved plans for Alabama, Florida, and South Carolina; see Gerald R. Webster, "Congressional Redistricting in the Southeastern U.S. in the 1990s," *Southeastern Geographer* 35 (1995): 1-21, esp. 5-6. For further information on preclearance, see Benjamin E. Griffith, "Annual Report of the Election Law and Reapportionment Subcommittees—Colorblind Jurisprudence: The Demise of Racial Rorschachism and Separatism," *Urban Lawyer* 30 (1998): 995-1022, esp. 996-1002.

63. Parker, *Black Votes Count,* 182-84.

64. Frank R. Parker, "The Mississippi Congressional Redistricting Case: A Case Study in Minority Vote Dilution," *Howard Law Journal* 28 (1985): 397-415. Parker includes a map of the state's 1981 plan (401).

65. *Jordan v. Winter,* 541 F. Supp. 1135, 1137 (N.D. Miss. 1982).

66. Parker, "Mississippi Congressional Redistricting," 408.

67. *Jordan v. Winter,* 604 F. Supp. 807, 814 (N.D. Miss. 1984). District populations varied narrowly, from 504,242 to 504,024. Later that year, the Supreme Court affirmed the district court's decision; see *Mississippi Republican Executive Committee v. Brooks,* 469 U.S. 1002 (1984).

68. McDonald, "1982 Amendments of Section 2 and Minority Representation," 76. Espy won nearly all of the black vote, 10 percent of the white vote, and 52 percent of the total vote.

CHAPTER THREE

1. Mississippi Republicans made only modest gains during the Reagan years. In 1980 the state's lower house included 116 Democrats, 4 Republicans, and 2 independents, while the upper house consisted of 48 Democrats and 4 Republicans. In 1990 Democrats outnumbered Republicans 112 to 9 in the lower house and 44 to 8 in the upper house. U.S. Bureau of the Census, *Statistical Abstract of the United States: 1981,* 102nd ed. (Washington, D.C., 1981), 494; and U.S. Bureau of the Census, *Statistical Abstract of the United States: 1996,* 116th ed. (Washington, D.C., 1996), 282.

2. "Redistricting in the States," *Congressional Quarterly Weekly Report* 49 (1991): 2717; "Redistricting in the States," *Congressional Quarterly Weekly Report* 49 (1991): 3324.

3. "Remap Movement in S.C., Miss.," *Congressional Quarterly Weekly Report* 49 (1991): 3658.

4. Charles Mahtesian, "Mississippi Sends New Map to Justice Department; North Carolina Legislature Remains Undecided Whether to Defend Plan or Redo It," *Congressional Quarterly Weekly Report* 50 (1992): 30; and "Redistricting in the States," *Congressional Quarterly Weekly Report* 50 (1992): 3357.

5. "Official 1990 Count by District," *Congressional Quarterly Weekly Report* 49 (1991): 1309-12.

6. Bob Benenson and others, "Partisans Try to Shape the Maps to Gain Control in '90s," *Congressional Quarterly Weekly Report* 49 (1991): 3696-3716, esp. 3708; and Mahtesian, "Mississippi Sends New Map."

7. District populations ranged from 515,314 to 513,853. See U.S. Bureau of the Census, *Population and Housing Characteristics for Congressional Districts of the 103rd Congress: Mississippi,* report no. 1990 CPH-4-26 (Washington, D.C., 1992), 1.

8. Ibid.

9. For discussion of lower participation rates among minority voters, see Alan Gartner, "Districting: A Second Vantage Point," *Cardoza Law Review* 14 (1993): 1277-86.

10. See, for example, Carol Swain, "Strategies for Increasing Black Representation in Congress," in *The Politics of Race: African Americans and the Political System,* ed. Theodore Rueter (Armond, N.Y.: Sharpe, 1995), 212-25, esp. 214-15.

11. *Ketchum v. Byrne,* 740 F. 2d 1398, 1415 (1984), cited by Arthur J. Anderson and William S. Dahlstrom, "Technological Gerrymandering: How Computers Can Be Used in the Redistricting Process to Comply with Judicial Criteria," *Urban Lawyer* 22 (winter 1990): 59-77, quotation on 70.

12. "The 67% Solution," *Fortune* 124 (August 26, 1991): 120.

13. Kimball Brace and others, "Minority Voting Equality: The 65 Percent Rule in Theory and Practice," *Law and Policy* 10 (1988): 43-62, quotation on 43.

14. Bernard Grofman and Lisa Handley, "Identifying and Remedying Racial Gerrymandering," *Journal of Law and Politics* 8 (1992): 345-404, esp. 380-81.

15. William M. Welch, "Activist Claims Win in Miss. Election," *USA Today,* April 14, 1993, 2A.

16. U.S. Bureau of the Census, *Statistical Abstract of the United States: 1996,* 275.

17. Juliana Gruenwald, "Statistics Stir the Passions of Parties' Boosters," *Congressional Quarterly Weekly Report* 55 (1997): 1434-36.

18. Robert Benenson and others, *Jigsaw Politics: Shaping the House after the 1990 Census* (Washington, D.C.: Congressional Quarterly, 1990), 103-5.

19. Beth Donovan, "North Carolina: A Common Interest," *Congressional Quarterly Weekly Report* 49 (1991): 3726.

20. Mahtesian, "Mississippi Sends New Map," 30; and Susan B. Glasser, "Ohio Dems Target Gradison; GOP Nixes Miller," *Roll Call,* February 10, 1992.

21. Charles Mahtesian and others, "North Carolina Tries Again; Georgia Map Rejected," *Congressional Quarterly Weekly Report* 50 (1992): 118.

22. See Susan B. Glasser, "Both Parties Call New N.C. Map 'Idiotic'; Democratic Legislature's Latest Redistricting Plan Includes Skinny 'I-85' District," *Roll Call,* January 30, 1992; Glasser, "Ohio Dems Target Gradison"; and Guy Gugliotta, "Redistricting Draws Computer Literate into Political Fray," *Washington Post,* February 25, 1992, A15.

23. Gugliotta, "Redistricting Draws Computer Literate into Political Fray."

24. Mark Monmonier, *Drawing the Line: Tales of Maps and Cartocontroversy* (New York: Henry Holt, 1995), 204.

25. For examples, see the North Carolina section of U.S. Bureau of the Census, *Congressional District Atlas: 103rd Congress of the United States* (Washington, D.C., 1993). On p. 10, in inset A for Cumberland County, District 8 surrounds a thin appendage of District 7, and on p. 20, in inset B for Forsyth County, District 10 surrounds polyplike extensions of District 5. Inset C, on p. 21, provides an even more intriguing example in which two parts of District 5 are connected at a point and one of these parts is connected in turn to the rest of the district by a short stretch of a stream.

26. For examples involving District 1, see the North Carolina section of U.S. Bureau of the Census, *Congressional District Atlas: 103rd Congress.* Thin corridors are readily apparent in inset A for Duplin and Wayne Counties on p. 16 and inset B for Granville and Vance Counties on p. 27.

27. As drafted with Aldus Freehand—I traced features on map images scanned from the atlas—congressional district lines are 0.7 points wide (0.01 inch; there are seventy-two points to an inch) and the map is at an approximate scale of fifty miles to 29.38 points (0.408 inch). Applying the resulting ratio scale of 1:7,800,000 (0.408/(50 × 5,280 × 12)) to the line width of 0.01 inch yields an equivalent ground width of 78,000 inches, or 1.2 miles (78,000/(12 × 5,280)).

28. James J. Kilpatrick, "N.C. Gerrymandering Is Ludicrous," *Chicago Sun-Times,* February 17, 1992, 19.

29. Glasser, "Both Parties Call New N.C. Map 'Idiotic.'"

30. For examples involving District 1, see the North Carolina section of U.S. Bureau of the Census, *Congressional District Atlas: 103rd Congress.* Instances of

point contiguity occur in inset A for Columbus County on p. 8 and inset A for Duplin and Wayne Counties on p. 16.

31. For discussion of small multiples, see Edward R. Tufte, *The Visual Display of Quantitative Information* (Cheshire, Conn.: Graphics Press, 1983), 170-75.

32. Land areas range from 829.3 square miles for District 12 to 8,150.6 square miles for District 1. Because district populations are nearly equal, these are the state's most and least densely populated congressional districts. See U.S. Bureau of the Census, *Population and Housing Characteristics for Congressional Districts of the 103rd Congress: North Carolina,* report no. 1990 CPH-4-35 (Washington, D.C., 1993), 17-18.

33. Van Denton, "House Democrats Offer District Plan; Map Similar to Congressional Proposal," *Raleigh News and Observer,* January 19, 1992, A1.

34. Seth Effron, "State's Redistricting Gets Federal Scrutiny," *Greensboro News and Record,* December 8, 1992, A1.

35. "Political Pornography," *Wall Street Journal,* September 9, 1991, A10.

36. "Redistricting in North Carolina Breaks Racial Lines," *All Things Considered,* May 3, 1992.

37. Ibid.

38. Van Denton, "Rehnquist Rejects GOP Effort to Block Congressional Races," *Raleigh News and Observer,* March 12, 1992, B1.

39. Richard Wolf, "N.C. District Fills the Bill—Sort Of," *USA Today,* February 5, 1992, 5A.

40. "Redistricting Error May Yet Be Acknowledged," *Greensboro News and Record,* December 9, 1992, A10.

41. "Let Sound Principles Shape New Districts," *Raleigh News and Observer,* January 23, 1992, A15.

42. Van Denton, "House, Senate Leaders Produce Similar Redistricting Proposals," *Raleigh News and Observer,* January 20, 1992, A1.

43. Joan Biskupic, "N.C. Case to Pose Test of Racial Redistricting," *Washington Post,* April 20, 1993, A4.

44. "Weld's Self-Serving Redistricting," *Boston Globe,* July 10, 1992, 22.

45. Denton, "Rehnquist Rejects GOP Effort."

46. Van Denton, "House OKs New Districts; Congressional Plan Has Two Black Seats," *Raleigh News and Observer,* January 24, 1992, B3.

47. Beyle is quoted in Ronald Smothers, "Congressional Races; Strangely Shaped Hybrid Creatures Highlight North Carolina's Primary," *New York Times,* April 30, 1992, 28.

48. U.S. Bureau of the Census, *Population and Housing Characteristics for Congressional Districts of the 103rd Congress: North Carolina,* 1-2.

49. Republicans also filed a lawsuit, which attacked the state's plan as a partisan gerrymander. On March 9, 1992, a panel of federal judges rejected the party's challenge, and two days later the Supreme Court refused to hear an appeal. See Charles Mahtesian, "Green Light to N.C. Races," *Congressional Quarterly Weekly Report* 50 (1992): 656.

50. Ronald Smothers, "Fair Play or Racial Gerrymandering? Justices Study a 'Serpentine' District," *New York Times*, April 16, 1993, B7.

51. *Shaw v. Barr*, 808 F. Supp. 461 (E.D. N.C. 1992); see Phil Duncan, "Court to Rule on Remap," *Congressional Quarterly Weekly Report* 50 (1992): 3822. The panel rendered an oral decision on April 27 and issued a formal written opinion on August 7.

52. *Shaw v. Reno*, 61 L.W. 4818 (1993). Also see Linda Greenhouse, "Court Questions Districts Drawn to Aid Minorities," *New York Times*, June 29, 1993, A1.

53. *Shaw v. Reno*, 4823.

54. "Political Pornography—II," *Wall Street Journal*, February 4, 1992, A14; *Shaw v. Reno*, 4820.

55. Political scientist Bernard Grofman, whom O'Connor cited, attributed the verse to "A. Wuffle, Assistant to Professor" at the University of California, whom he interviewed in April 1992. See Bernard Grofman, "Would Vince Lombardi Have Been Right If He Had Said: 'When It Comes to Redistricting, Race Isn't Everything, It's the *Only* Thing'?" *Cardozo Law Review* 96 (1993): 1237-76, quotation on 1261. Also see *Shaw v. Reno*, 4820.

56. *Shaw v. Reno*, 4821.

57. Ibid., 4826.

58. Ibid.

59. "U.S. District Court Upholds 'Gerrymander' for Blacks," *New York Times*, August 3, 1994, A12.

60. Peter Applebome, "Suits Challenging Redrawn Districts That Help Blacks," *New York Times*, February 14, 1994, A1.

61. Ronald Smothers, "Black District in Georgia Is Ruled Invalid," *New York Times*, September 13, 1994, A14. For the comment on shape, see Applebome, "Suits Challenging Redrawn Districts."

62. For a news report, see Kevin Sack, "North Carolina in Turmoil over Redistricting Ruling," *New York Times*, June 15, 1996, 6. For the majority opinion and a dissenting opinion by Justice Stevens, see *Shaw v. Hunt*, 116 Sup. Ct. 1894 (1996).

63. *Shaw v. Hunt*, 1899.

64. Ibid.

65. Ibid., 1906. The internal quotations are Rehnquist's references to language used in earlier cases.

66. Ibid., 1915.

67. Ibid., 1907.

68. Ibid., 1922. Joining Stevens were Justices Ruth Bader Ginsburg and Stephen Breyer. Justice David Souter, whom Justices Ginsburg and Breyer also joined, referred to his dissent in another recent case, *Bush v. Vera*.

69. *Bush v. Vera*, 116 Sup. Ct. 1941 (1996).

70. *Miller v. Johnson*, 115 Sup. Ct. 2475, 2490 (1995). For discussion of the lower court's ruling, see Smothers, "Black District in Georgia Is Ruled Invalid," A14.

71. See Ronald D. Elving, "Court Ban on Race-Based Maps Keeps District Lines in Flux," *Congressional Quarterly Weekly Report* 55 (1997): 752; Alan Greenblatt, "Texas, North Carolina Maps Upheld by Federal Panels," *Congres-*

sional Quarterly Weekly Report 55 (1997): 2250; and "Virginia Legislature Revisiting House District Map," *Congressional Quarterly Weekly Report* 56 (1998): 140. Although a district court panel had ordered Louisiana to redraw its congressional districts, the Supreme Court overturned the lower-court order because the plaintiffs did not live in the black-majority district in question. See *U.S. v. Hayes,* 115 Sup. Ct. 2431 (1995).

72. Sack, "North Carolina in Turmoil."

73. Dennis Patterson, "Congressional Districts: Some Say Proposal Has No Minority-Majority Districts," *Wilmington Morning Star,* July 25, 1996, 3B.

74. Ibid.

75. "North Carolina Gets Redistricting Reprieve," *New York Times,* August 22, 1996, A21.

76. Joseph Neff and Joe Dew, "Ruling Puts off Redistricting until 1997," *Raleigh News and Observer,* July 31, 1996, A1.

77. "North Carolina's Map Heads to Justice, Judges for OK," *Congressional Quarterly Weekly Report* 55 (1997): 810.

78. Wade Rawlins, "Panel OKs Redrawn Districts," *Raleigh News and Observer,* September 16, 1997, A3.

79. "District 12 Ruling Has Hopefuls in Limbo," *Greensboro News and Record,* April 15, 1998, B1; Geoff Earle, "Federal Panel Throws a Curve to North Carolina with Remap," *Congressional Quarterly Weekly Report* 56 (1998): 947; and "Ruling May Force a Second Primary," *Greensboro News and Record,* April 5, 1998, A1.

80. "NC12: State Defends Redrawn 12th CD," *Bulletin's Forerunner,* June 3, 1998. Also see Geoff Earle, "North Carolina Redistricting Plan Advances," *Congressional Quarterly Weekly Report* 56 (1998): 1384.

81. "North Carolina House Map Clears," *Roll Call,* June 15, 1998; and Geoff Earle, "Judges Approve North Carolina Redistricting," *Congressional Quarterly Weekly Report* 56 (1998): 1751.

82. *Hunt v. Cromartie,* 119 Sup. Ct. 1545 (1999). Also see Caroline E. Brown, "High Court Upholds Minority Districts," *Congressional Quarterly Weekly Report* 57 (1999): 1202; and Linda Greenhouse, "Court Gives Wiggle Room to Racially Drawn Districts," *New York Times,* May 18, 1999, A21. Also see Linda Greenhouse, "A Fight on Redistricting Returns to the High Court," *New York Times,* January 21, 1999, A14.

83. *Hunt v. Cromartie,* 1551.

84. Ibid.

85. As the North Carolina General Assembly Web site noted, "A Supreme Court decision has since reversed [the lower-court ruling that led to the 1998 map]. Barring further court action, '97 HOUSE/SENATE PLAN A' will be used in the 2000 elections." See <http://www.ncga.state.nc.us/html1999/geography/html4Trans/District_Files/Congress/97HSPlanA/97HAS_Main.html > (August 9, 1999).

86. Elving, "Court Ban on Race-Based Maps Keeps District Lines in Flux," 752.

87. *King v. State Board of Elections,* 979 F. Supp. 582, 617 (N.D. Ill. 1996).

88. *King v. Illinois Board of Elections,* 118 Sup. Ct. 877 (1998); and "High Court OKs Illinois 4th," *Congressional Quarterly Weekly Report* 56 (1998): 249.

89. *King v. State Board of Elections,* 587.

CHAPTER FOUR

1. Quoted in Alan Greenblatt, "Texas, North Carolina Maps Upheld by Federal Panels," *Congressional Quarterly Weekly Report* 55 (1997): 2250. Redistricting officials in Georgia and Louisiana had also removed plaintiffs from challenged districts in those states.

2. The North Carolina General Assembly provided summary statistics and maps at its Web site, at <http://www.ncga.state.nc.us> (July 24, 1998).

3. Associated Press, "State's District Plan Is Defended in Brief," *Raleigh News Observer,* June 2, 1998; retrieved from <http://www.news-observer.com> Web site (June 19, 1998).

4. U.S. district court data on the black percentage of the voting-age population were published in Julina Gruenwald, "Judges' Georgia Map Divides Black-Majority Districts," *Congressional Quarterly Weekly Report* 53 (1995): 3837-38.

5. Kevin Sack, "A Redistricted Black Lawmaker Fires Back," *New York Times,* July 4, 1996, B6; Kevin Sack, "Lawmakers Survive First Tests," *New York Times,* July 11, 1996, A14; and Kevin Sack, " Victory of 5 Redistricted Blacks Recasts Gerrymandering Dispute," *New York Times,* November 23, 1996, 1.

6. Gruenwald, "Judges' Georgia Map Divides Black-Majority Districts."

7. For insight on Republican gains in the South, see Christopher Caldwell, "The Southern Captivity of the GOP," *Atlantic Monthly* 281 (June 1998): 55-72.

8. Deborah Kalb, "Party-Switchers' Fate Tough to Predict," *Congressional Quarterly Weekly Report* 54 (1996): 1001.

9. "Black- and Hispanic-Majority Districts," *Congressional Quarterly Weekly Report* 51 (1993): 1829.

10. Kenneth J. Cooper, "Election Plan Toes the Line in Louisiana," *Washington Post,* April 16, 1994, A4. Also see "Ruling Leads to Action in Louisiana Case," *Congressional Quarterly Weekly Report* 51 (1993): 1829.

11. Dave Kaplan, "Majority-Black District Rejected in Louisiana," *Congressional Quarterly Weekly Report* 52 (1994): 2175.

12. Joan Biskupic, "Race-Conscious District Reinstated in Louisiana," *Washington Post,* August 12, 1994, A3; and "Supreme Court Allows Use of Louisiana Map in 1994," *Congressional Quarterly Weekly Report* 52 (1994): 2373.

13. Dave Kaplan, "Incumbents Make a Clean Sweep," *Congressional Quarterly Weekly Report* 52 (1994): 2911.

14. *United States v. Hays* and *Louisiana v. Hays,* 115 Sup. Ct. 2431 (1995).

15. Juliana Gruenwald, "Revised Challenge Filed in Louisiana Case," *Congressional Quarterly Weekly Report* 53 (1995): 2097.

16. *Miller v. Johnson,* 115 Sup. Ct. 2475, 2486 (1995). In *Miller,* the Supreme Court upheld a decision of the federal district court.

17. Ibid., 2488.

18. "Minority Districts," *Congressional Quarterly Almanac* 51 (1995): 6-39 to 6-40, quotation on 6-39.

19. For a small-scale version of the court's 1994 map, see Kaplan, "Majority-Black District Rejected."

20. John McQuaid and Ed Anderson, "La.'s Black Majority Districts Now Down to One," *New Orleans Times-Picayune,* June 25, 1996, A6; and Tim Curran, "High Court Refuses Redistricting Appeal," *Roll Call,* June 27, 1996. By a 8-1 majority, the justices refused to hear the appeal.

21. Tim Curran, "Redistricting Fault Lines Swallow Victims," *Roll Call,* July 15, 1996.

22. *Bush v. Vera,* 116 Sup. Ct. 1941, 1955 (1996). Originally filed as *Vera v. Bush,* the case was on appeal from the district court, which had overturned the three districts.

23. Ibid., 1959.

24. Ibid., 1983.

25. Ibid.

26. Ibid., 1982-83.

27. Ibid., 1958.

28. For discussion of the ramifications of "strict scrutiny," see Jennifer R. Abrams, "The Supreme Court's Disenfranchisement of the American Electorate: Advocating the Application of Strict Scrutiny When Reviewing State Ballot Access Laws and Political Gerrymandering," *St. John's Journal of Legal Commentary* 12 (1996): 145-69; and David P. Swartz, "Constitutional Law—Strict Scrutiny Applies to Congressionally Mandated Race-Conscious Programs—Adarand Constructors, Inc. v. Pena, 115 Sup. Ct. 2097 (1995)," *Suffolk University Law Review* 30 (1996): 277-84.

29. See Ian Ayres, "Narrow Tailoring," *UCLA Law Review* 43 (1996): 1781-838; and William J. Crowley, "*Miller v. Johnson* (115 Sup. Ct. 2475 (1996)) and the Case for Compliance with the Voting Rights Act as a Compelling State Interest," *George Mason University Civil Rights Law Journal* 6 (1996): 65-95.

30. *King v. Illinois Board of Elections,* 118 Sup. Ct. 877 (1998).

31. Carl Redman, "Department Rejects La.'s Remap Plan," *Baton Rouge Advocate,* August 14, 1996, 1A.

32. *Abrahms v. Johnson,* 117 Sup. Ct. 1925 (1997). Also see Linda Greenhouse, "U.S. Loses Redistricting Challenge," *New York Times,* June 20, 1997, A14.

33. *Reno v. Bossier Parish School Board,* 117 Sup. Ct. 1491 (1997). For analysis, see Associated Press, "Reaction Mixed over Ruling on Minority Districts," *St. Louis Post-Dispatch,* May 13, 1997, 9A; John Aloysius Farrell, "Supreme Court Hinders Black Majority Districts," *New Orleans Times-Picayune,* May 13, 1997, A1; and Linda Greenhouse, "Curb on U.S. Role in Redistricting," *New York Times,* May 13, 1997, A16. Also see Linda Greenhouse, "Justices Say Redistricting Need Only Prevent Backsliding," *New York Times,* January 25, 2000, A18.

34. Jonathan P. Hicks, "Albany Lawmakers Agree on Redrawn 12th District," *New York Times,* July 31, 1997, B3; and Greenblatt, "Texas, North Carolina Maps Upheld by Federal Panels."

35. Bill Tracking Web site, 1998 session, Virginia General Assembly, at <http://www.leg1.state.va.us> (July 28, 1998). Senate Bill 13 describes the remap, which changed five of Virginia's eleven congressional districts. Also see Tyler Whitley, "Justice Department Clears State's Redistricting Plan," *Richmond Times Dispatch,* March 13, 1998, A6.

CHAPTER FIVE

1. For comparative evaluations of compactness indexes, see Mark S. Flaherty and William W. Crumplin, "Compactness and Electoral Boundary Adjustment: An Assessment of Alternative Measures," *Canadian Geographer* 36 (1992): 159-71; David L. Horn, Charles R. Hampton, and Anthony J. Vandenberg, "Practical Application of District Compactness," *Political Geography* 12 (1993): 103-20; and Alan M. MacEachren, "Compactness of Geographic Shape: Comparison and Evaluation of Measures," *Geografiska Annaler* 67B (1985): 53-67. For an essay advocating numerical standards, see Daniel D. Polsby and Robert D. Popper, "The Third Criterion: Compactness as a Procedural Safeguard against Partisan Gerrymandering," *Yale Law and Policy Review* 9 (1991): 301-53. For a more recent approach to coding and representing shape, see John K. Wildgen, "Fractal Geometry and the Boundaries of Voting Districts," in *Spatial and Contextual Models in Political Research,* ed. Munroe Eagles (London: Taylor and Francis, 1995), 107-25.

2. Richard H. Pildes and Richard G. Niemi, "Expressive Harms, 'Bizarre Districts,' and Voting Rights: Evaluating Election-District Appearances after *Shaw v. Reno,*" *Michigan Law Review* 92 (1993): 483-587.

3. Alan MacEachren, who applied eleven compactness measures to a representative sample of fifty-four United States counties, observed high to moderate correlation among the indexes. Correlation coefficients ranged from 0.70 to 1.00, and most were 0.85 or higher. The dispersion and perimeter scores, labeled indexes C and B in MacEachren's study, yielded a correlation coefficient of 0.76 and were among the least redundant pairs. See MacEachren, "Compactness of Geographic Shape."

4. Pildes and Niemi, "Expressive Harms, 'Bizarre Districts,' and Voting Rights," 571-73.

5. Florida's District 3 linked urban neighborhoods in Gainesville, Ocala, Lake City, Orlando, and Jacksonville. Struck down by a federal three-judge panel in April 1996, the district was redrawn the following month by the state legislature, which apparently was well aware of both the inevitability of the district court's ruling and the futility of appealing to the Supreme Court. See Juliana Gruenwald, "Florida Black-Majority District Is Ruled Unconstitutional," *Congressional Quarterly Weekly Report* 54 (1996): 1071; and Juliana Gruenwald, "Florida Lawmakers Agree on New Map," *Congressional Quarterly Weekly Report* 54 (1996): 1241.

6. The study reports each state's minimum scores. If we assume that Georgia's Eleventh and Virginia's Third Districts had the least efficient boundaries within the state—a reasonable conclusion at least for the former—these districts just

missed the cutoff with perimeter scores of 0.07 and 0.06, respectively. See Pildes and Niemi, "Expressive Harms, 'Bizarre Districts,' and Voting Rights," 571-73.

7. Ibid., 587.

8. Despite numerous caveats, compactness measures can be useful for comparing different plans within a state. And individual states can, as two have done, impose their own formulas and constraints. In Colorado, for example, the state constitution requires that "each district shall be as compact in area as possible and the aggregate linear distance of all district boundaries shall be as short as possible." See West's C.R.S.A. Const. Art. 5, §47. And Iowa's state election law includes two standards for compactness: the ratio of a district's length and width and the ratio of dispersion of population around the population center to the dispersion of population around the geographic center. See §42.4 in Iowa General Assembly, *Code of Iowa, 1991, Containing All Statutes of a General and Permanent Nature* (Des Moines, 1990), 365-66. In the twenty-three other states with a constitutional or statutory compactness requirement, numerical measures have no official status, except when introduced at a trial by an expert witness. According to Pildes and Niemi, who examined legal challenges to noncompact districts, state courts are reluctant to strike down contorted districts. Compactness is little more than an abstract goal, and without numerical guidelines that make violations readily apparent, state courts hesitate to interfere with the political process. Not surprisingly, Colorado and Iowa are among the few states in which the courts have enforced the compactness requirement. See Pildes and Niemi, "Expressive Harms, 'Bizarre Districts,' and Voting Rights," 527-31.

9. Pildes and Niemi, "Expressive Harms, 'Bizarre Districts,' and Voting Rights," 537.

10. *Bush v. Vera,* 116 Sup. Ct. 1941, 1958 (1996).

11. Michael Barone and Grant Ujifusa, *Almanac of American Politics* (Washington, D.C.: National Journal, 1996), 1335, cited in *Bush v. Vera,* 116 Sup. Ct. 1941, 1958 (1996).

12. *Bush v. Vera,* 116 Sup. Ct. 1941, 1982 (1996).

13. For further insight on the concept of *functional compactness,* see Richard K. Scher, Jon L. Mills, and John J. Hotaling, *Voting Rights and Democracy: The Law and Politics of Districting* (Chicago: Nelson-Hall, 1997), 152-53. Functional compactness was apparently important in the Court's rejection of a challenge to California's post-1990 remap; see *DeWitt v. Wilson,* 856 F. Supp. 1409 (E.D. Cal. 1994) and 515 U.S. 1170 (1995). Three legal scholars recommend California's post-1990 strategy as a guide for Texas and other large states with substantial minority populations; see David M. Guinn, Christopher W. Chapman, and Kathryn S. Knechtel, "Redistricting in 2001 and Beyond: Navigating the Narrow Channel between the Equal Protection Clause and the Voting Rights Act," *Baylor Law Review* 51 (1999): 225-67. For a challenge to the validity of geometric compactness as a criterion for redistricting, see Christian R. Grose, "Do Compact Congressional Disticts Enhance Constituent Knowledge and Communication?" paper presented at GIS and Political Redistricting: Social Groups, Representational Values, and Electoral Boundaries, a conference sponsored by the National Center

for Geographic Information and Analysis and held in late October 1997 at the State University of New York at Buffalo.

14. Remark attributed to House candidate Mickey Michaux in Susan B. Glasser, "Both Parties Call New N.C. Map 'Idiotic'; Democratic Legislature's Latest Redistricting Plan Includes Skinny 'I-85' District," *Roll Call,* January 30, 1992.

15. The delta resembles a bird's foot because of multiple talonlike distributary channels, which the river developed and later abandoned as the delta extended itself forward into the Gulf of Mexico. Geomorphologists consider the Mississippi the prototypical bird's-foot delta. See, for example, O. D. von Engeln, *Geomorphology: Systematic and Regional* (New York: Macmillan, 1948), 254.

16. Joseph Schwartzberg, a geographer and an early advocate of numerical measurement of the compactness of voting districts, noted in the 1960s that a state's boundary may make compactness difficult to achieve. He cited Cape Cod and Long Island as examples and opined that "only a dozen or so" congressional districts are "affected to a significant degree by the shape of the state." See Joseph E. Schwartzberg, "Reapportionment, Gerrymanders, and the Notion of 'Compactness,'" *Minnesota Law Review* 50 (1966): 443-52, quotation on 448.

17. For examples, see Robert B. McMaster, "Automated Line Generalization," *Cartographica* 24 (summer 1987): 74-111, and Robert B. McMaster and K. Stuart Shea, *Generalization in Digital Cartography* (Washington, D.C.: Association of American Geographers, 1992).

18. The more generalized eighteen- and thirty-four-point examples in figure 5.5 still exhibit substantial curvature because the boundary is reconstructed as a series of smooth mathematical curves known as Bezier splines.

19. See, for example, Horn, Hampton, and Vandenberg, "Practical Application of District Compactness"; Richard G. Niemi, Bernard Grofman, Carl Carlucci, and Thomas Hofeller, "Measuring Compactness and the Role of a Compactness Test for Partisan and Racial Gerrymandering," *Journal of Politics* 52 (1990): 1155-81; Ruth C. Silva, "Reapportionment and Redistricting," *Scientific American* 213 (November 1965): 20-27; and James B. Weaver and Sidney W. Hess, "A Procedure for Nonpartisan Districting: Development of Computer Techniques," *Yale Law Journal* 73 (1963): 288-308.

20. For an analysis using population-comparison scores based on the convex hull, see Niemi, Grofman, Carlucci, and Hofeller, "Measuring Compactness and the Role of a Compactness Standard," esp. 1163 and 1174-76.

21. For a fuller discussion, see Weaver and Hess, "Procedure for Nonpartisan Districting," 296-300.

22. Horn, Hampton, and Vandenberg, "Practical Application of District Compactness," 107.

23. Pildes and Niemi, "Expressive Harms, 'Bizarre Districts,' and Voting Rights," 557.

24. Andrew Gelman and Gary King, "Enhancing Democracy through Legislative Redistricting," *American Political Science Review* 88 (1994): 541-59; and

Micah Altman, "Districting Principles and Democratic Representation" (Ph.D. diss., California Institute of Technology, 1988), 3.

25. For examples, see David Butler and Bruce Cain, *Congressional Redistricting: Comparative and Theoretical Perspectives* (New York: Macmillan, 1992), 148-51; Bruce Cain, "Perspectives on *Davis v. Bandemer:* Views of the Practitioner, Theorist, and Reformer," in *Political Gerrymandering and the Courts,* ed. Bernard Grofman (New York: Agathon Press, 1990), 117-42; Robert G. Dixon Jr., "The Warren Court Crusade for the Holy Grail of 'One Man-One Vote,'" in *Supreme Court Review, 1969,* ed. Philip B. Kurland (Chicago: University of Chicago Press, 1969), 219-70; and Daniel Hays Lowenstein, "Bandemer's Gap: Gerrymandering and Equal Protection," in *Political Gerrymandering and the Courts,* ed. Bernard Grofman (New York: Agathon Press, 1990), 64-116.

CHAPTER SIX

1. For further discussion of these themes, see Mark Monmonier, *How to Lie with Maps,* 2nd ed. (Chicago: University of Chicago Press, 1996). For a brief but insightful account of the persuasive power of graphics, see Kris Goodfellow, "Pictures Speak Much Louder Than Words in Courtroom Animations," *New York Times,* November 24, 1997, D1.

2. For a fuller description of the hearing, held on February 1, 1992, at the Onondaga County Court House, see Mark Monmonier, *Drawing the Line: Tales of Maps and Cartocontroversy* (New York: Henry Holt, 1995), 189-93.

3. Nominating petitions and absentee ballots are often thrown out for inconsequential technicalities, the budget is several months late in most years, legislators abhor a rainy-day fund but delight in funding pork-barrel "member items," and—for a state one might expect to be progressive—public referenda are extremely difficult to place on the ballot. For insight on politics and government in New York, see Jeffrey M. Stonecash, John Kenneth White, and Peter W. Colby, eds., *Governing New York State,* 3rd ed. (Albany: State University of New York Press, 1994).

4. New York State Legislative Task Force on Demographic Research and Reapportionment, *Co-chairmen's Proposed 1992 Assembly and State Senate District Boundaries, January 21, 1992,* 3.

5. Ibid., 9.

6. New York State Legislative Task Force on Demographic Research and Reapportionment, *1992 State Senate and Assembly Districts Approved by the Legislature, April 2, 1992.*

7. New York State Legislative Task Force on Demographic Research and Reapportionment, *Co-chairmen's Proposed 1992 Assembly and State Senate District Boundaries,* 1.

8. Ibid., 2.

9. "Cartographic silences" is an elegant and powerful term promoted by historian of cartography Brian Harley (1932-91), who studied the suppression of facts or features on the official maps of governments eager to assert territorial claims or champion the rights of property owners. See, for example, J. B. Harley, "Silences and Secrecy: The Hidden Agenda of Cartography in Early Modern

Europe," *Imago Mundi* 40 (1988): 57-76; and J. B. Harley, "Cartography, Ethics, and Social Theory," *Cartographica* 27 (summer 1990): 1-23.

10. Kenneth C. Martis, *The Historical Atlas of United States Congressional Districts, 1789-1983* (New York: Free Press, 1982).

11. See, for example, the before-and-after maps that accompany Jonathan P. Hicks, "Road Gets Tougher for a Political Pioneer," *New York Times*, February 9, 1998, B3.

12. For examples of the use of cartoonlike maps to communicate contempt, see Paul V. Niemeyer, "The Gerrymander: A Journalistic Catch-Word or Constitutional Principle? The Case in Maryland," *Maryland Law Review* 54 (1995): 242-60.

13. Deval L. Patrick, "What's Up Is Down, What's Black Is White," *Emory Law Journal* 44 (1995): 827-45, quotation on 832.

14. Ibid., 833.

15. Ibid.

16. Quoted in Peter Applebome, "Key Trial on House Districts to Begin," *New York Times*, March 28, 1994, A11.

17. *Bush v. Vera*, 116 Sup. Ct. 1941, 1958 (1996).

18. John Hart Ely, "Gerrymanders: The Good, the Bad, and the Ugly," *Stanford Law Review* 50 (1998): 607-41, quotation on 619.

19. Ibid., 611-14, quotations on 611 and 607.

20. Also see Daniel D. Polsby and Robert D. Popper, "The Third Criterion: Compactness as a Procedural Safeguard against Partisan Gerrymandering," *Yale Law and Policy Review* 9 (1991): 301-53.

21. Hampton Dellinger, "Words Are Enough: The Troublesome Use of Photographs, Maps, and Other Images in Supreme Court Opinions," *Harvard Law Review* 110 (1997): 1704-53, quotation on 1710.

22. The other states are Alabama, California, and Pennsylvania. For the maps, see *Colgrove v. Green*, 328 U.S. 549, 560-63 (1946).

23. For the map of New Jersey congressional districts, see *Karcher v. Daggett*, 77 L. Ed. 2d 133, 150.

24. *Karcher v. Daggett*, 462 U.S. 725, 755 (1983).

25. Ibid., 787.

26. *Davis v. Bandemer*, 478 U.S. 109, 173 (1986), quotation in note 12.

27. *Shaw v. Reno*, 509 U.S. 630, 633 (1993).

28. Ibid., 647.

29. *Shaw v. Hunt*, 116 Sup. Ct. 1894, 1899 (1986).

30. Dellinger, "Words Are Enough," esp. 1752-53, quotation on 1742.

31. Ibid., 1753.

CHAPTER SEVEN

1. For a concise overview of historical trends toward an ever greater advantage for incumbents, see Gary W. Cox and Jonathan N. Katz, "The Reapportionment Revolution and Bias in U.S. Congressional Elections," *American Journal of Political Science* 43 (1999): 812-40.

2. John T. McQuiston, "Samuel S. Stratton, 73; Lawmaker Represented Albany in Congress," *New York Times,* September 15, 1990, 29.

3. Bill Becker, "House Seats Lost by G.O.P. Upstate," *New York Times,* November 5, 1958, 30. Stratton's victory was also the lead in an Associated Press story published in Syracuse; see "New York Democrats Gain 2 House Seats," *Syracuse Post-Standard,* November 5, 1958, 2.

4. Warren Weaver Jr., "Stratton Winner in Upstate Voting," *New York Times,* November 7, 1962, 1; also see "Stratton Wins over Gordon in Big Upset," *Syracuse Post-Standard,* November 7, 1962, 2.

5. Weaver, "Stratton Winner in Upstate Voting," 17.

6. Leroy Hardy, *The Gerrymander: Origin, Conception and Re-emergence* (Claremont, Calif.: Rose Institute of State and Local Government, 1990), 29-32.

7. Ibid., 31.

8. Ibid., 30-31.

9. U.S. Congress, House of Representatives, *Statistics of the Presidential and Congressional Election of November 4, 1952* (Washington, D.C., 1953), 3.

10. Mark Monmonier, *Drawing the Line: Tales of Maps and Cartocontroversy* (New York: Henry Holt, 1995), 209-12.

11. See Bob Benenson, "N.Y. Legislators Pass New Map but It Faces Legal Challenge," *Congressional Quarterly Weekly Report* 50 (1992): 1733; Kevin Sack, "Albany Legislators Agree on Plan for Revised Congressional Lines, *New York Times,* June 4, 1992, A1; and Kevin Sack, "Redistricting Plans Approved," *New York Times,* July 3, 1992, B1.

12. Solarz, who is Jewish, had hoped the five Hispanic candidates would split the ethnic vote. Although he promised, if elected, to move to the district, Latino leaders denounced the nine-term congressman as a carpetbagger and an opportunist. Press accounts suggest that Solarz's district was dismembered because he was "one of the most frequent offenders in the House bank scandal." See Lindsey Gruson, "Solarz Will Run in District Tailored as a Hispanic Seat," *New York Times,* July 9, 1992, B3. Also see Alison Mitchell, "Rep. Solarz Loses in a New District," *New York Times,* September 16, 1992, A1.

13. Fred P. Graham, "High Court Voids State Districts, Orders New Plan," *New York Times,* April 8, 1969, 1.

14. With Republicans controlling the governorship and the state senate, Democrats lacked the clout for vindictive redistricting—but so did the Republicans. The principal differences between the 1962 and 1968 plans were in New York City and its suburbs. Editions of *Congressional District Atlas* published in 1966 and 1968 indicate that the 1968 remap affected only two upstate counties, recombining one that had been split and reassigning the other. The 1969 redistricting was inherently more brutal: riding Nixon's coattails, the GOP recaptured the assembly in 1968 and openly applauded the opportunity afforded by the Supreme Court ruling. Graham, "High Court Voids State Districts." Also see Peter Kihss, "G.O.P. Districting May Help Rivals," *New York Times,* April 9, 1969, 30; Sydney H. Schanberg, "State Reshapes House Districts; Court Fight Due," *New*

York Times, February 27, 1968, 1; and Sidney E. Zion, "State Republicans See a Gain of 6 to 8 House Seats," *New York Times,* April 8, 1969, 34.

15. Edward Ranzal, "U.S. Court Orders State to Equalize Districts for '70," *New York Times,* June 19, 1969, 39. The one person, one vote doctrine and population gains in New York City suburbs also had a profound effect on the state legislature; see Richard C. Lehne, Guthrie Birkhead, and H. George Frederickson, "New York Reapportionment: The City and the State," *Maxwell Review* 8 (winter 1971-72): 27-32.

16. Richard Reeves, "State Remapping of Congress Lines Helpful to G.O.P.," *New York Times,* January 20, 1970, 1. The new plan had a disparity of only 409 persons.

17. Richard L. Madden, "Battle of 2 Incumbents Gives State Unusual Race," *New York Times,* October 1, 1970, 37.

18. Ibid.

19. " . . . Westchester, Upstate," *New York Times,* October 23, 1970, 40.

20. Richard L. Madden, "Lowenstein Loses Seat in Congress," *New York Times,* November 4, 1970, 1.

21. Frank Lynn, "Republicans Add Bipartisan Touch to Redistricting," *New York Times,* December 18, 1971, 19; and Richard L. Madden, "The Season for Accommodation," *New York Times,* December 18, 1971, 19.

22. Richard L. Madden, "Peyser-Ottinger Result Still in Doubt," *New York Times,* November 9, 1972, 26; and Richard L. Madden, "Reid Wins as Democrat; Bella Abzug Easy Victor," *New York Times,* November 8, 1972, 1.

23. Alan Hevesi, "The Renewed Legislature," in *New York State Today: Politics, Government, Public Policy,* ed. Peter W. Colby (Albany: State University of New York Press, 1985), 147-60, esp. 151-52.

24. Stratton apparently moved during his first term as representative for the new Twenty-third District. The *Official Congressional Directory* for the Ninety-eighth Congress, 1983-84, reports Stratton's residence as Amsterdam. Subsequent editions list his residence as Schenectady.

25. For a concise statement of the Task Force's role and organization, see George A. Mitchell, ed., *The New York Red Book,* 93rd ed. (Guilderland: New York Legal Publishing, 1995), 397.

26. Although a federal district court judge issued an order prohibiting the Census Bureau from delivering the final count to President Carter, the Supreme Court quickly voted seven to one to overturn the injunction. The case apparently was dropped because even a plausible adjustment would not have been sufficient to save one of the lost seats. See E. J. Dionne Jr., "Census Study Finds City Losing Legislative Seats," *New York Times,* August 23, 1979, B1; Arnold H. Lubasch, "Census Bureau Told to Readjust Figures for City and State," *New York Times,* December 24, 1980, A1; Irvin Molotsky, "High Court Lets Census Officials Give Data Today," *New York Times,* December 31, 1980, B3; and Alan Murray, "Census Bureau Presents Final Figures," *Congressional Quarterly Weekly Report* 39 (1981): 4.

27. Penelope E. Harvison, Robert C. Speaker, and Marshall L. Turner Jr.,

"Drawing the Lines—By the Numbers: The Statistical Foundations of the Electoral Process," *Government Information Quarterly* 2 (1985): 389-405.

28. Nancy Meyer, "Redistricting: Redrawing the Lines," *Empire State Report* 16 (November 1990): 11-14; and Kevin Sack, "The Great Incumbency Machine," *New York Times Magazine,* September 27, 1992, 47-62, esp. 54. For insight on earlier reapportionments, dominated largely by Republicans, see David I. Wells, "The Reapportionment Game," *Empire State Report* 5 (February 1979): 8-14; and David I. Wells, "Redistricting in New York State: It's a Question of Slicing the Salami," *Empire* 4 (October–November 1978): 9-13.

29. U.S. Bureau of the Census, *Statistical Abstract of the United States, 1982-83,* 103rd ed. (Washington, D.C., 1982), 487; and U.S. Bureau of the Census, *Statistical Abstract of the United States, 1992,* 112th ed. (Washington, D.C., 1992), 266.

30. Sack, "Great Incumbency Machine," 48. According to Julian Palmer, director of New York State Common Cause, "Since 1986, state legislators have been more likely to be indicted than to lose a general election."

31. In April 1990, for a study of the use of geographic information systems (GIS) in state government in New York, I visited Task Force headquarters in lower Manhattan, a short walk from New York City Hall. At that time there were about thirty people on the staff, plus a few others in Albany. Preparing for the decennial remap, the staff was bit larger than a few years earlier, when the Task Force worked on projects for various committees of the legislature as well as the Department of Environmental Conservation, Metro North, the State Banking Association, and the Solid Waste Management Commission. For the 1980 remap, staff had worked three shifts; for 1990 at least two shifts seemed inevitable.

The office had recently converted to Arc/Info software and belonged to "ReDNet," a group of fourteen states planning to use the software for redistricting. Personnel I interviewed had three concerns for the post-1990 remap: population equality; the Department of Justice, which had to approve the legislative and congressional plans; and incumbents, who did not want to run against each other or in a new district where they were not well known.

I also visited the group in April 1992 while collecting information for *Drawing the Line.* The legislature had recently agreed on assembly and senatorial districts, and the Task Force was working on the state's formal submission to the Department of Justice. While the plans were still under development, staff in Albany would link the maps to statistical models of voting behavior, to assess how well the plans would perform. At the time of my visit legislators had yet to compromise on the congressional remap.

32. The diskettes contained recent voting data and census tabulations but lacked an electronic block-level geographic base map, available only on computer tape. See New York State Legislative Task Force on Demographic Research and Reapportionment, *Public Access Procedures* (New York, 1990).

33. Ibid., 3.

34. Monmonier, *Drawing the Line,* 209-12; and Kevin Sack, "Albany Getting an Earful over Congressional Map," *New York Times,* March 10, 1992, A1.

35. Kevin Sack, "Two Methods to Redistrict Sent to Court," *New York Times,* April 28, 1992, B2.

36. Associated Press, "Redistricting Battle Has Become a War," *Syracuse Post-Standard,* June 12, 1992, A8; Benenson, "N.Y. Legislators Pass New Map," 1733; Kevin Sack, "Court Backs New Plan for Districts," *New York Times,* June 10, 1992, B6; Kevin Sack, "Court May Impose Redistricting Plan," *New York Times,* June 12, 1992, B4; Kevin Sack, "Cuomo Still Pondering Signing or Vetoing Districting Bill," *New York Times,* May 4, 1992, B5; Todd S. Purdum, "New Court Order Heightens Confusion on Redistricting," *New York Times,* May 14, 1992, B8; and Sam Howe Verhovek, "New York Court Throws a State Plan into Redistricting Fray," *New York Times,* June 2, 1992, B1.

37. Sack, "Redistricting Plans Approved," B1.

38. Robert Benenson and others, *Jigsaw Politics: Shaping the House after the 1990 Census* (Washington, D.C.: Congressional Quarterly, 1990), 29-31. In the early 1990s, thirty-three states followed the textbook "how a bill becomes law" formula, as did nine others with significant quirks. With a permanent bipartisan agency appointed by and responsible to legislative leaders, New York was one of the latter. *Congressional Quarterly*'s editors did not consider the Task Force an independent commission because it draws up—under the direct influence of legislative leaders—an initial plan that is then treated like other legislation.

39. For further discussion, see Jeffrey C. Kubin, "The Case for Redistricting Commissions," *Texas Law Review* 75 (1997): 837-72.

40. Although Montana empowered its commission to draw up the initial plan for congressional districts, slow growth during the 1980s cost the state one of its two House seats. Figure 7.3 does not reflect two other developments. A federal court declared Michigan's commission unconstitutional before the group could redraw boundaries for the state's post-1990 remap, and Idaho established a commission that will produce its first plans for the post-2000 remap.

41. Ibid., 862-63.

42. Ibid., 865.

43. Rex Dean Honey and Douglas Deane Jones, "An Ethical Escape from the Political Thicket: The Iowa Model for Electoral Districting," *Papers and Proceedings of the Applied Geography Conferences* 14 (1991): 92-99; and Edward Walsh, "Redistricting without Politics: Iowa's Lines Are Drawn by a Disinterested Party," *Washington Post,* May 10, 1991, A3. Also see *Iowa Annotated Code,* §2.58-65 and §42.1-6 (West 1993). A permanent state agency with a full-time director, general counsel, and assorted staff, the Legislative Service Bureau is advised by the Temporary Redistricting Advisory Commission. The majority and minority floor leaders of the two houses each appoint a member of the commission, and these four members elect a fifth member to serve as chair.

44. If counties or cities must be split, the bureau must make a greater effort to preserve the boundaries of units with smaller populations. By law, compactness is measured two ways: the length-width ratio of a district's longest axis to its shorter perpendicular axis, and the distance between a district's geometric centroid (center of area) and a centroid weighted according to the populations of constituent

small areas (center of population). Honey and Jones, "Ethical Escape from the Political Thicket," 94-95.

45. Dan Balz and Edward Walsh, "Lines Drawn in Iowa," *Washington Post,* May 14, 1991, A5; and Rhodes Cook, "Iowa Remapping Goes Smoothly as Six Districts Become Five," *Congressional Quarterly Weekly Report* 49 (1991): 1306-8. Wyoming, which has a single seat in the House, had completed its legislative remap before Iowa.

46. Cook, "Iowa Remapping Goes Smoothly," quotation on 1306. Although forced to run against a Republican incumbent, Nagle called the congressional remap "a fair plan." Despite a slight advantage in voter registrations, Nagle lost to incumbent Republican Jim Nussle. For a concise overview of Iowa's plans for the next remap, see "Preparations for the Year 2000 Census: Iowa's Role for ArcView GIS in Redistricting," *ESRI ARC News* 21 (spring 1999): 25.

47. All provinces and the Northwest Territories have a boundary commission. Yukon Territory, with a single seat, has no intraprovincial boundaries to readjust.

48. Richard W. Jenkins, "Untangling the Politics of Electoral Boundaries in Canada, 1993-1997," *American Review of Canadian Studies* 28 (1998): 517-38. Except in the most extraordinary circumstances, ridings may not differ from the provincial average ("quota") population by more than 25 percent.

49. Despite the protest, a survey of MPs' attitudes suggests that Canada's boundary commissions are working well; see Munroe Eagles and R. Kenneth Carty, "MPs and Electoral Redistribution Controversies in Canada, 1993-96," *Journal of Legislative Studies* 5 (1999): 74-95.

50. D. Butler, "The Redrawing of Parliamentary Boundaries in Britain," *Journal of Behavioral and Social Sciences* 37 (1992): 5-12.

51. D. J. Rossiter, R. J. Johnston, and C. J. Pattie, "Estimating the Partisan Impact of Redistricting in Great Britain," *British Journal of Political Science* 27 (1997): 319-31.

52. Johnston and his coworkers attribute Labour's victory in the 1997 election to two factors: the party's more effective presentation to the boundary commissions and a campaign focused on winnable seats; see R. J. Johnston, C. J. Pattie, and David Rossiter, "Anatomy of a Labour Landslide: The Constituency System and the 1997 General Elections," *Parliamentary Affairs* 51 (1998): 131-48.

53. For a critique of "neutral criteria" and "neutral procedures," see David Butler and Bruce Cain, *Congressional Redistricting: Comparative and Theoretical Perspectives* (New York: Macmillan, 1992), 145-51. For a vigorous denunciation of "neutral public interest criteria" and "nonpartisan commissions," see Daniel H. Lowenstein and Jonathan Steinberg, "The Quest for Legislative Districting in the Public Interest: Elusive or Illusory?" *UCLA Law Review* 33 (1985): 1-76.

CHAPTER EIGHT

1. William Vickrey, "On the Prevention of Gerrymandering," *Political Science Quarterly* 76 (1961): 105-10, quotations on 105 and 110.

2. Curtis C. Harris Jr., "A Scientific Method of Districting," *Behavioral Science* 9 (1964): 219-25, quotations on 219.

3. James B. Weaver and Sidney W. Hess, "A Procedure for Nonpartisan Districting: Development of Computer Techniques," *Yale Law Journal* 73 (1963): 288-308, quotations on 308.

4. Technological determinism, which has "hard" and "soft" variants, is more appropriately treated as "technological momentum." For insights on this debate, see Robert L. Heilbroner, "Do Machines Make History?" and Thomas P. Hughes, "Technological Momentum," in *Does Technology Drive History? The Dilemma of Technological Determinism,* ed. Merritt Roe Smith and Leo Marx (Cambridge: MIT Press, 1994), 53-65 and 101-13.

5. Donald F. Cooke, "Topology and TIGER: The Census Bureau's Contribution," in *The History of Geographic Information Systems: Perspectives from the Pioneers,* ed. Timothy W. Foresman (Upper Saddle River, N.J.: Prentice Hall, 1998), 47-57; esp. 51-52. As a mathematician in the Census Bureau's Statistical Research Division, Corbett was attempting to expedite the bureau's development of a database of block and tract boundaries. In devising a file structure that identified the cobounding 2-cells sharing a common boundary, he simplified coding and "data capture" by allowing each shared boundary to be digitized (measured and recorded) once, not twice.

6. Nonstreet boundaries were possible before TIGER, but DIME required the partition of curved boundaries into a series of short straight-line segments and the cumbersome addition of extra links and nodes. Timothy F. Trainor, "Fully Automated Cartography: A Major Transition at the Census Bureau," *Cartography and Geographic Information Systems* 17 (1990): 27-38, esp. 33-34. Also see Frederick R. Broome and David B. Meixler, "The TIGER Data Base Structure," *Cartography and Geographic Information Systems* 17 (1990): 39–47.

7. Michelle H. Browdy, "Simulated Annealing: An Improved Computer Model for Political Redistricting," *Yale Law and Policy Review* 8 (1990): 163-79.

8. For further information on location-allocation modeling for legislative redistricting, see Richard L. Morrill, *Political Redistricting and Geographic Theory,* Resource Publications in Geography (Washington, D.C.: Association of American Geographers, 1981), 36–42.

9. Weaver and Hess, "Procedure for Nonpartisan Districting."

10. For discussion of other approaches to automated redistricting see Edward Forrest, "Apportionment by Computer," *American Behavioral Scientist* 8 (December 1964): 23, 35; Stuart S. Nagel, "Simplified Bipartisan Computer Districting," *Stanford Law Review* 17 (1965): 863-99; Terry B. O'Rourke, *Reapportionment: Law, Politics, Computers* (Washington, D.C.: American Enterprise Institute for Public Policy Research, 1972), 73-97; and Ruth C. Silva, "Reapportionment and Redistricting," *Scientific American* 213 (November 1965): 20-27, esp. 26-27. For a contemporary critique, see Micah Altman, "The Computational Complexity of Automated Redistricting: Is Automation the Answer?" *Rutgers Computer and Technology Law Journal* 23 (1997): 81-142.

11. Morrill, *Political Redistricting,* 53-61.

12. Richard L. Morrill, "Redistricting Revisited," *Annals of the Association of American Geographers* 66 (1976): 548-56.

13. Table 8.1 is adapted from ibid., 552. Although the seven district values for the straight-line solution sum to 105,799, my version lists the lower figure (105,304) reported by Morrill.

14. Table 8.2 is adapted from ibid., 553. Although travel-time measurements reflect the sum of each election district's population multiplied by the average travel time from the election distict's center to the hypothetical center of the congressional district, Morrill's article does not specify the units reported.

15. Ibid., 555-56.

16. According to Brian Hayes, a former editor of the *American Scientist,* "a few state legislative districts were drawn by computer programs in Iowa in 1967 and in Delaware in 1968. As far as I know, computer-automated redistricting has not been given a serious trial since then." Brian Hayes, "Machine Politics," *American Scientist* 84 (November–December 1996): 522-26, quotation on 524. Writing in the early 1980s, political historian Gordon Baker attributed "ever more elaborate gerrymanders in many states" following the 1970 and 1980 censuses to computers, which "brought a degree of sophistication to boundary manipulation," but his passing reference says nothing about algorithmic strategies. See Gordon E. Baker, "Representation and Apportionment," in *Encyclopedia of American Political History: Studies of the Principal Movements and Ideas,* ed. Jack P. Greene (New York: Charles Scribner's Sons, 1984), 1118-30, quotations on 1129.

17. See, for example, John M. Littschwager, "The Iowa Redistricting System," *Annals of the New York Academy of Science* 219 (1973): 221-35.

18. See, for example, Arthur J. Anderson and William S. Dahlstrom, "Technological Gerrymandering: How Computers Can Be Used in the Redistricting Process to Comply with Judicial Criteria," *Urban Lawyer* 22 (1990): 59-77; David L. Anderson, "When Restraint Requires Action: Partisan Gerrymandering and the Status Quo Ante," *Stanford Law Review* 42 (1990): 1549-76; and Michelle H. Browdy, "Computer Models and Post-*Bandemer* Redistricting," *Yale Law Journal* 99 (1990): 1379-98. For a more pessimistic appraisal, see Samuel Issacharoff, "Judging Politics: The Elusive Quest for Judicial Review of Political Fairness," *Texas Law Review* 71 (1993): 1643-1703.

19. Critics of automated redistricting also argue that because the results tend to confer benefits on a particular candidate or party, a computerized remap is hardly neutral; for examples, see Robert G. Dixon Jr., *Democratic Representation: Reapportionment in Law and Politics* (New York: Oxford University Press, 1968), 532-35; and Anderson and Dahlstrom, "Technological Gerrymandering."

20. Altman, "Computational Complexity of Automated Redistricting," quotations on 107 and 134.

21. Redistricting consultants, who now advertise on the World Wide Web, have advised many state party organizations. No consultant is better known than Kimball Brace, whom *Congressional Quarterly* called "a pioneer in the use of computers for redistricting" and a "secret weapon" for Illinois Democrats. In the post-1980 Illinois remap, Brace's firm, Election Data Services, identified flaws in the Republicans' plans and helped transform the Democrats' 86-91 minority in the legislature's lower house into a 70-48 majority. See Robert Benenson and oth-

ers, *Jigsaw Politics: Shaping the House after the 1990 Census* (Washington, D.C.: Congressional Quarterly, 1990), 73-76, quotations on 73.

22. Ibid., 136.

23. For an overview of the role of geographic information systems in redistricting, see Munroe Eagles, Richard S. Katz, and David Mark, "GIS and Redistricting," *Social Science Computer Review* 17 (1999): 5-9. For examples of software systems and consultants' services, see Gary H. Anthes, "GIS Eases Redistricting Worry," *Computer World,* October 7, 1991, 65; Charles Lindauer, "Tactical Cartography," *Campaigns and Elections* 20 (April 1999): 48-51; William E. Schmidt, "New Age of Gerrymandering: Political Magic by Computer?" *New York Times,* January 10, 1989, A1; and Mark Thompson, "The Gerrymanderer's Dream Machine," *California Lawyer* 10 (January 1990): 20-22. Investment in a GIS also held postredistricting benefits for organizations eager to improve the efficiency of their efforts at advertising, campaigning, and getting out the vote; see Winnett W. Hagens and Anthony E. Fairfax, "Precision Voter Targeting: GIS Maps out a Strategy," *Geo Info Systems* 6 (November 1996): 24-30.

24. *Johnson v. Miller,* 864 F. Supp. 1354, 1362 (S.D. Ga. 1994).

25. Ibid., 1361.

26. Ibid., 1363.

27. Ibid.

28. *Vera v. Richards,* 861 F. Supp. 1304, 1318 (S.D. Tex. 1994).

29. Ibid., 1319.

30. *Shaw v. Hunt,* 861 F. Supp. 408 (E.D. N.C. 1994), quotations on 408.

31. Ibid., 457.

CHAPTER NINE

1. For examples, see Sue Garrison, "Wariness, Weariness and Other Worries of the 2000 Census," *Business Geographics* 6 (November 1998): 42-43; Steven A. Holmes, "Republicans Resisting Census Plan to Correct Undercounting," *New York Times,* May 27, 1997, A13; and Rochelle L. Stanfield, "Sermons on the Count," *National Journal* 30 (August 8, 1998): 1854-60. The mandate for a census is a sentence in Article 1, Section 2, of the United States Constitution: "The actual Enumeration shall be made within three Years after the first Meeting of the Congress of the United States, and within every subsequent Term of ten Years, in such Manner as they shall by Law direct."

2. Linda Greenhouse, "In Blow to Democrats, Court Says Census Must Be by Actual Count," *New York Times,* January 26, 1999, A1; and Charles Pope, "Supreme Court Ruling Offers Little Hope for Ending Census Sampling Dispute," *Congressional Quarterly Weekly Report* 318 (1999): 259-60.

3. For a concise overview of the 1999 controversy, see Margo Anderson and Stephen E. Fienberg, "To Sample or Not to Sample? The 2000 Census Controversy," *Journal of Interdisciplinary History* 30 (summer 1999): 1-36; David Baumann, "Still No Consensus on the Census," *National Journal* 31 (1999): 956-57. For an account of the 1995 shutdown, see Fran Durbin, "Agencies Report Furloughs, Lost Sales in Six-Day Shutdown," *Travel Weekly* 54 (November 27, 1995):

1; Richard Morin, "Public Sides with Clinton in Fiscal Fight," *Washington Post*, November 21, 1995, A4; and "The Shutdown beyond the Beltway: Budget Bill on Verge of Passage and Veto," *Minneapolis Star-Tribune*, November 18, 1995, 1A.

4. Illegal aliens are to be counted like everyone else. In 1979, the Federation for American Immigration Reform sued to have the Census Bureau exclude illegal aliens from the reapportionment data. A federal court, upheld on appeal, dismissed the suit because the issue had been decided in the 1920s, when Congress debated but rejected a proposal not to count illegal aliens. Margo Anderson, *The American Census: A Social History* (New Haven: Yale University Press, 1988), 229.

5. Some observers argue that the effects of adjustment are so minor it's not worth the effort and annoyance. See, for example, Peter Skerry, "Sampling Error," *New Republic* 220 (May 31, 1999): 18-20.

6. Undercount rates for 1990 are from U.S. General Accounting Office, *2000 Census: Progress Made on Design, but Risks Remain*, report GGD-97-142 (Washington, D.C., July 1997), 7-8.

7. For a history of the undercount and efforts to improve coverage, see Margo A. Conk, "The 1980 Census in Historical Perspective," in *The Politics of Numbers*, ed. William Alonso and Paul Starr (New York: Russell Sage Foundation, 1987), 155-86. For insights on the political realities of census enumeration and statistical adjustment, see Margo Anderson and Stephen E. Fienberg, "Who Counts? The Politics of Censustaking," *Society* 34 (March–April 1997): 19-26; and Paul Starr, "The Sociology of Official Statistics," in Alonso and Starr, *Politics of Numbers*, 7-57.

8. Starr, "Sociology of Official Statistics," 7.

9. College students should be counted where they attend school, whereas elementary and high-school students attending boarding school are counted at their parents' homes. Michael R. Lavin, *Understanding the Census: A Guide for Marketers, Planners, Grant Writers, and Other Data Users* (Kenmore, N.Y.: Epoch Books, 1996), 115.

10. Harvey M. Choldin, *Looking for the Last Percent: The Controversy over Census Undercounts* (New Brunswick, N.J.: Rutgers University Press, 1994), 210; and Panel on Census Requirements in the Year 2000 and Beyond, Committee on National Statistics, Commission on Behavioral and Social Sciences and Education, National Research Council, *Modernizing the U.S. Census* (Washington, D.C.: National Academy Press, 1995), 33-34.

11. Stephen Magagnini, "The Big Count: Local Governments Are Revving up for Some Changes in Census 2000," *Planning* 64 (May 1998): 13-16, quotation on 13.

12. Lavin, *Understanding the Census*, 81.

13. Ibid., 71-73. For the 1990 census, the Census Bureau relied heavily on lists purchased from direct marketing firms.

14. Ibid., 9.

15. Ibid.

16. "Native Americans Pose Census Challenge," *Syracuse Post-Standard*, April 16, 1998, A6.

17. Magagnini, "Big Count."

18. For a description and critique of "S-night," see Choldin, *Looking for the Last Percent,* 195-99; and Lavin, *Understanding the Census,* 85. S-Night, on March 20-21, was followed on March 31 by Transient Night, when census workers canvassed labor camps, circuses, carnivals, and other sites with transient populations. For examples of the fears and frustrations of census workers, see Gwen Ifill and Barbara Vobejda, "Counting Homeless Proves Daunting Task for Census Bureau," *Washington Post,* March 22, 1990, A3.

19. See, for example, "Homeless Tally in Los Angeles Called Flawed," *Washington Post,* June 10, 1990, A10. By contrast, an editorial in the *National Review* viewed the Census Bureau's admittedly conservative estimate as evidence that anti-Reagan ideologues had exaggerated the homeless problem; see "Whither the Homeless? Census Bureau Report Shows Fewer Homeless Persons Than Expected," *National Review* 43 (May 13, 1991): 18.

20. Guy Gugliotta, "Institute Finds a Number That Adds up, Has Meaning on the Streets," *Washington Post,* May 16, 1994, A3.

21. Panel to Evaluate Alternative Census Methods, Committee on National Statistics, Commission on Behavioral and Social Sciences, National Research Council, *A Census That Mirrors America: Interim Report* (Washington, D.C.: National Academy Press, 1993), 25.

22. U.S. General Accounting Office, *2000 Census,* 5. For a concise overview of the Census Bureau's extensive outreach and promotional activities, see Lavin, *Understanding the Census,* 54-55, 75-76, 88-89.

23. The General Accounting Office accepted the Census Bureau's estimate that sampling for nonresponse follow-up could trim the cost of the 2000 census by as much as $800 million; see U.S. General Accounting Office, *2000 Census,* 73.

24. For an explanation of the logic underlying the Census Bureau's proposed sampling procedure, see Tommy Wright, "Sampling and Census 2000: The Concepts," *American Scientist* 86 (1998): 245-53; Tommy Wright and Howard Hogan, "CENSUS 2000: Evolution of the Revised Plan," *Chance* 12, 4 (1999): 11-19. For additional insights, see Harvey M. Choldin, "How Sampling Will Help Defeat the Undercount," *Society* 34 (March–April 1997): 27-30; David Kestenbaum, "Census 2000: Where Science and Politics Count Equally," *Science* 279 (1998): 798-99; Nathan Keyfitz, "The Case for Census Tradition," *Society* 34 (March–April 1997): 45-48; Ivars Peterson, "Sampling and the Census: Improving the Accuracy of the Decennial Count," *Science News* 152 (1997): 238-39; and Gordon F. Sutton, "Is the Undercount a Demographic Problem?" *Society* 34 (March–April 1997): 31-35.

25. Magagnini, "Big Count," quotation on 14. Also see Steven A. Holmes, "Panels Move to Bar Data Sampling for Census," *New York Times,* September 18, 1996, A14.

26. Holmes, "Republicans Resisting Census Plan to Correct Undercounting," A13.

27. David Broder, "GOP Minds Made up on Census," *Syracuse Post-Standard,* August 20, 1997, A12; Jerry Gray, "Flood Relief Bill Passes as G.O.P., in

Turmoil, Yields," *New York Times,* June 13, 1997, A1; and Jerry Gray, "In Spending Bill, Gauntlet on Census Is Thrown Down," *New York Times,* October 1, 1997, A23.

28. Dan Carney, "Commerce-Justice Funding Goes down to the Wire," *Congressional Quarterly Weekly Report* 55 (1997): 2855-56; Jerry Gray, "Senate Adjusts Legislation, Hoping to Remove Obstacles and Avoid Veto," *New York Times,* November 10, 1997, A28; and Steven A. Holmes, "Tentative Pact Will Allow Census to Test the Sampling Method," *New York Times,* November 1, 1997, A12. The Clinton administration and Republican leaders extended the compromise the following year, while the Republicans' case moved through lower federal courts on its way to the Supreme Court; see Steven A. Holmes, "Fight over Statistical Sampling in 2000 Is Postponed," *New York Times,* October 16, 1998, A24.

29. "Truce on Census Sampling Reached," *Population Today* 26 (January 1998): 5. The Sacramento and Menominee tests used traditional as well as sampling strategies, while a third dress rehearsal, in counties surrounding Columbia, South Carolina, tested only a traditional, direct-enumeration strategy. Sacramento posed the challenge of an ethnically and racially diverse population, with many newly arrived minority residents, while the Menominee site posed the challenges of idiosyncratic addresses and Native American resistance to government.

30. Randolph E. Schmid, "Census Reports Successful Dress Rehearsal," Associated Press Wire, July 9, 1998. Letter carriers delivered the forms in Sacramento, and census workers dropped off forms on the Menominee reservation.

31. John Howard, "Counts Differ as Federal Government Unveils Dress Rehearsal," Associated Press State and Local Wire, January 15, 1999.

32. Press releases apparently listed the statistically adjusted population as 4,738, but a later report raised it to 4,779 people. See "Census Test in Menominee County a Success," *Wisconsin State Journal,* July 9, 1998, 4C; Robert Imrie, "Census Says Dress Rehearsal Misses 143 People in Menominee County," Associated Press State and Local Wire, February 25, 1999; and U.S. Bureau of the Census, *Census 2000 Dress Rehearsal: Public Law 94-171 Product Documentation* (Washington, D.C., 1999), table 1.

33. Greenhouse, "In Blow to Democrats, Court Says Census Must Be by Actual Count," A1; and Pope, "Supreme Court Ruling Offers Little Hope," 259-60.

34. *Department of Commerce v. House of Representatives,* 142 L. Ed. 2d 797 (1999), quotation on 820.

35. Ibid., 819.

36. Ibid.

37. Steven A. Holmes, "Census Ruling Is Said to Cost $1.7 Billion," *New York Times,* June 3, 1999, A20; Steven A. Holmes, "Ruling Said to Raise Census Cost by $2 Billion," *New York Times,* February 24, 1999, A18; Dick Kirschten, "Politicizing the Census," *Government Executive* 31 (March 1999): 14; and U.S. Bureau of the Census, *Updated Summary: Census 2000 Operational Plan* (February 1999). In a classic example of Washington's shameless manipulation of words and numbers, congressional budget leaders chose to treat the 2000 census—a predictable

decennial occurrence—as an "emergency" appropriation; see "The Census as an Act of God" [editorial], *New York Times,* July 29, 1999, A20.

38. U.S. Bureau of the Census, *Updated Summary,* 1.

39. Lavin, *Understanding the Census,* 358.

40. U.S. Bureau of the Census, *Updated Summary: Census 2000 Operational Plan Using Traditional Census-Taking Methods* (January 1999) XI-5.

41. John H. Cushman Jr., "Census Bills Advance in House Despite the Bureau's Protests," *New York Times,* March 18, 1999, A19; and Jack W. Germond and Jules Witcover, "When Addition Begets Division," *National Journal* 31 (1999): 909.

42. Robert Pear, "House Approves Bill to Allow Local Challenges of 2000 Census," *New York Times,* April 14, 1999, A19. Local review, administration supporters argued, would interfere with sampling by adding nine weeks to the process and threatening the Census Bureau's ability to meet its deadline for delivering redistricting data; see Baumann, "Still No Consensus on the Census."

43. For a mathematical introduction to imputation, see Donald B. Rubin, *Multiple Imputation for Nonresponse in Surveys* (New York: John Wiley, 1987), esp. 9. For an explanation and concise overview of substitution and allocation in processing the 1990 census, see Lavin, *Understanding the Census,* 358–64. A third kind of imputation, the *confidentiality edit,* preserves the confidentiality of households and individuals by injecting error into published statistics. Induced uncertainty is especially important for areas with small populations because unaltered tabulations might, for example, reveal too much about the income reported by a particularly wealthy individual.

44. This is called the "last resort method." Lavin, *Understanding the Census,* 358.

45. Logic also plays a role in allocation, as when no gender—Census Bureau staff members still say "sex"—is reported for a child identified as a daughter.

46. These rates, like those mapped in figure 9.4, are from table 284 in U.S. Bureau of the Census, *1990 Census of Population: General Population Characteristics, United States,* report CP-1-1 (Washington, D.C., November 1992), 589.

47. Ibid., 588.

48. Choldin, *Looking for the Last Percent,* 46–65. In 1980 the Census Bureau inaugurated a more ambitious postenumeration survey, which matched a sample of 70,000 households with a second sample drawn from 100,000 completed census forms. The strategy is similar conceptually to the capture-recapture surveys with which biologists estimate wildlife populations. For example, ornithologists wanting to estimate the number of purple finches in an area might capture, band, and release one thousand of the birds. A short time later, they observe that one-twentieth of the purple finches captured in a second canvass have bands from the first survey. If they assume the two samples were drawn randomly from the same population, it follows that the population of purple finches is twenty times larger than the one thousand individuals in first sample. For human populations, census officials increase the reliability of their estimates by partitioning census tracts and blocks into *strata* based on demographic and socioeconomic characteristics and carrying out separate estimates for each stratum.

The Census Bureau also uses sampling to extrapolate to the entire population information gleaned from its more detailed "long form" questionnaire, received by 17 percent of households in 1990. Because sampling works, tabulations based on the long form are published for census tracts as well as larger areal units. See Lavin, *Understanding the Census,* 9, 12-13, 346-54.

49. Anderson and Fienberg, "Who Counts?" For discussion of the 1990 postenumeration survey and proposals for adjusting the reapportionment and redistricting populations, see Choldin, *Looking for the Last Percent,* 206-26; Lavin, *Understanding the Census,* 355-58; and Panel on Census Requirements in the Year 2000 and Beyond, *Modernizing the U.S. Census,* 85-112. For a critique of the panel's report, see House Committee on Post Office and Civil Service, *Review of Interim Report by the National Academy of Sciences on Census Reform: Hearing before the Subcommittee on Census, Statistics, and Postal Personnel,* 103rd Cong., 1st sess., 1993.

50. One example is the Be Counted program. To "provide a means for people to be included in Census 2000 if they believe they have not received a census questionnaire or were not included on one," the Census Bureau developed its Be Counted program to distribute forms printed in Chinese, English, Korean, Spanish, Tagalog, and Vietnamese at "Walk-in Questionnaire Assistance Centers." See U.S. Bureau of the Census, *Updated Summary, Census 2000 Operational Plan Using Traditional Census-Taking Methods,* IX-3. And as in previous censuses, a deluge of advertisements and public service announcements urged everyone who could read or hear to "be counted."

51. Rochelle L. Stanfield, "Multiple Choice," *National Journal* 29 (1997): 2352-55. The son of a black father and a Thai mother, Tiger Woods recognizes white and American Indian blood in his ancestry. Activists alarmed that multiracial categories might dilute the clout of blacks criticize his reluctance to declare himself African American. Gregory P. Kane, "Push for 'Multiracial' Label Gives Rise to Suspicions among Blacks," *Tampa Tribune,* July 18, 1997, 13.

52. Lavin, *Understanding the Census,* 138-39.

53. For a social scientist's critique of the "continuing salience of race in American Society" perpetrated by the federal statistics system, see Kenneth Prewitt, "Public Statistics and Democratic Politics," in *The Politics of Numbers,* ed. William Alonso and Paul Starr (New York: Russell Sage Foundation, 1987), 261-74, quotation on 271-72. In 1998, President Clinton appointed Prewitt director of the Census Bureau.

54. I estimated the rate as 2.0 percent from rates of 1.9 and 2.2 percent listed for 1991 and 1992, respectively, in the table "Race of Wife by Race of Husband," which does not report an interracial marriage rate for 1990. The table's estimates for 1991 and 1992 are from the Current Population Survey, not the census. See table 1 at U.S. Census Bureau, "Interracial Tables," at <http://www.census.gov/population/www/socdemo/interrace.html> (May 26, 1999).

55. The 2.9 percent rate is based on the 1990 census. Ibid., table 2; and "Census Race and Ethnic Categories Retooled," *Population Today* 26 (January 1998): 4.

56. Eric Schmitt, “Experts Clash over Need for Changing Census Data by Race,” *New York Times,* April 24, 1997, A27.

57. A federal task force assigned to study racial and ethnic designations announced its recommendations on July 8, 1997, and on October 29, after a brief comment period and some minor modifications, the Office of Management and Budget adopted the new classification. See Steven A. Holmes, “Panel Balks at Multiracial Census Category,” *New York Times,* July 9, 1997, A12; Steven A. Holmes, “People Can Claim One or More Races on Federal Forms,” *New York Times,* October 30, 1997, A1; and William O’Hare, “Managing Multiple-Race Data,” *American Demographics* 20 (April 1998): 42–44.

58. Linda Mathews, “More Than Identity Rides on a New Racial Category,” *New York Times,* July 6, 1996, 1.

59. Christine B. Hickman, “The Devil and the One Drop Rule: Racial Categories, African Americans, and the U.S. Census,” *Michigan Law Review* 95 (1997): 1161-265, quotation on 1255.

60. See U.S. Census Bureau, “Findings on Questions of Race and Hispanic Origin Tested in the 1996 National Content Survey,” Population Division Working Paper 16 (Washington, D.C., 1997), available at <http://www.census.gov/population/www/documentation/twps0016> (May 28, 1999).

61. Because OMB did not give ethnicity the flexibility accorded race, a person with a single Hispanic parent has but one opportunity, like everyone else, to affirm or deny a Spanish-Hispanic-Latino identity. And despite the check-box convenience of declaring oneself Cuban, Mexican, or Puerto Rican, the multiorigin offspring of a Cuban American and a Mexican American, for instance, can select only one ethnic homeland.

62. For a concise explanation, see Joel Perlmann, “Reflecting the Changing Face of America,” Jerome Levy Economics Institute, Bard College, Public Policy Brief, no. 35A (September 1997), highlights, at <http://www.levy.org/docs/hili35a.html> (May 24, 1999). Also see Christy Fisher, “It’s All in the Details,” *American Demographics* 20 (April 1998): 46–47.

63. Fisher, “It’s All in the Details.”

64. Tabulation Working Group, Interagency Committee for the Review of Standards for Data on Race and Ethnicity, “Draft Provision Guidance on the Implementation of the 1997 Standards for Federal Data on Race and Ethnicity,” February 17, 1999.

65. The groups listed here are reported in various parts of the “PL 20 Matrix” described in ibid., 20-22, 128-34. As of late July 1998, the Justice Department had registered no objection to the PL 20 tabulations, which the Census Bureau used for reporting sample redistricting data for the 1998 dress rehearsal. The following month Marshall Turner, chief of the Census 2000 Redistricting Data Office, noted that “the Voting Rights Section reviewed this smaller PL 20 Matrix and in late July informed the Census Bureau that this product would meet the census information needs associated with Sections 2 and 5 of the Voting Rights Act.” For additional details on the Census Bureau’s effort to seek consensus on racial/ethnic tabulations, see Marshall L. Turner, “Dress Rehearsal Prototype Redistrict-

ing Data," memo dated August 24, 1998, at <http://www.census.gov/clo/www/plmemo3.htm> (March 8, 1999). A more complex treaatment of race is still possible: on April 6, 2000, the Census Bureau announced a more flexible, more detailed format with sixty-three separate racial categories, which users could (if those chose) collapse to the six or seven listed here. Marshall L. Turner, e-mail communication, June 13, 2000. Also see Steven A. Holmes, "New Policy on Census Says Those Listed as White and Minority Will Be Counted as Minority," *New York Times,* March 11, 2000, A9.

66. Ibid., 42.

67. Panel on Census Requirements for the Year 2000 and Beyond, *Modernizing the U.S. Census,* 38-40.

68. Whatever mathematical dissonance is apparent reflects the January 1999 Supreme Court ruling and a law that preceded it. In deciding *Department of Commerce v. House of Representatives,* the high court ruled that the state totals, which the Census Bureau must deliver by December 31, 2000, and which are used to apportion the House of Representatives, will not reflect corrections based on the Census Bureau's Accuracy and Coverage Evaluation survey. By contrast, the redistricting data, which the bureau must deliver by April 1, 2001, will correct for undercounts and overcounts. But Public Law 105-119, a 1997 statute, required that the bureau also release the redistricting data without corrections. Marshall Turner, e-mail communication, June 2, 1999.

69. For a taste of the alarmist attacks likely to bombard the nation's op-ed pages, see Grover Norquist, "Census Bureau's Secret," *Washington Times,* June 1, 1999, A12. A byline identified Norquist as the president of Americans for Tax Reform and chairman of Citizens for an Honest Count Coalition. A key concern was a "manipulated census" that, in the name of adjusting for the overcount, would delete "millions of Americans who don't own two homes and didn't send in two census forms." Norquist conveniently overlooked the fact that statistical adjustment, in attempting to make the count accord with reality, never fingers individual households or specific persons.

CHAPTER TEN

1. "Lani Guinier Named U.S. Civil Rights Chief; 1st Black Woman in Post," *Jet* 84 (May 17, 1993): 5; Paul Gigot, "Hillary's Choice on Civil Rights: Back to the Future," *Wall Street Journal,* May 7, 1993, A14; and Ordine E. Le Blanc, "Lani Guinier," *Contemporary Black Biography* (Detroit: Gale Research, 1994), 7:107-12.

2. See, for example, Clint Bolick, "Clinton's Quota Queens," *Wall Street Journal,* April 30, 1993, A12. For examples of her law review articles, see Lani Guinier, "Groups, Representation, and Race-Conscious Districting: A Case of the Emperor's Clothes," *Texas Law Review* 71 (1993): 1589-642; Lani Guinier, "The Representation of Minority Interests: The Question of Single-Member Districts," *Cardozo Law Review* 14 (1993): 1135-74; and Lani Guinier, "The Triumph of Tokenism: The Voting Rights Act and the Theory of Black Electoral Success," *Michigan Law Review* 89 (1991): 1077-1154. A year after Clinton had

withdrawn her nomination, she published several of them as chapters in Lani Guinier, *The Tyranny of the Majority: Fundamental Fairness in Representative Democracy* (New York: Free Press, 1994).

3. White House officials asked Guinier not to discuss controversial issues publicly until her confirmation hearing. The president announced his withdrawal of the nomination on June 3, 1993. In a short press conference following the announcement, he claimed to be unaware of her views: "At the time of the nomination, I had not read her writings. In retrospect, I wish I had. Today, as a matter of fairness to her, I read some of them again in good detail. They clearly lend themselves to interpretations that do not represent the views that I expressed on civil rights during my campaign, and views that I hold very dearly, even though there is much in them with which I agree." See "Transcript of President Clinton's Announcement," *New York Times,* June 4, 1993, A19. For the nominee's recollections, see Lani Guinier, *Lift Every Voice: Turning a Civil Rights Setback into a New Vision of Social Justice* (New York: Simon and Schuster, 1998), esp. 114–31.

4. Kenneth C. Martis, *The Historical Atlas of United States Congressional Districts, 1789–1983* (New York: Free Press, 1982).

5. Ibid., 4–5, quotation on 4.

6. Ibid., 72, 249, 263. In the early 1830s, Philadelphia occupied a small part of Philadelphia County.

7. Ibid., 4.

8. The law mandates contiguity and single-member districts in a single sentence: "elected by districts composed of contiguous territory equal in number to the representatives to which the said state may be entitled." Ibid.

9. In particular, see Guinier, *Tyranny of the Majority,* 14–16.

10. For a history and description of the Cambridge system, see the Proportional Representation Web site of the Intelligent Information Infrastructure Project, in the Artificial Intelligence Laboratory at Massachusetts Institute of Technology, at <http://www.ai.mit.edu/projects/iiip/Cambridge/prop-voting/prop-voting.html> (June 11, 1999). For additional discussion of the Cambridge voting method, see Howard Fain, "P.R. Elections in Cambridge, Mass.," *National Civic Review* 83 (1994): 84–85. The city council elects one of its members to serve as the mayor, who presides at council meetings, chairs the school committee, and serves as the city's chief ceremonial officer. Day-to-day city operations are directed by the city manager, elected by the council, which also appoints the city clerk and the city auditor.

11. According to Cambridge elections official Teresa Neighbor, the lower-ranked candidates are then promoted: "For example, if the voter gives a #1 to candidate A, a #2 to candidates B and C, and a #3 to candidate D, the #3 for the last candidate is promoted to a #2, when and if that ballot is transferred. If the voter gives a #1 to candidates A and B, and a #2 to candidate C, then the #2 for candidate C is promoted to #1." E-mail communication with Teresa Neighbor, June 25, 1999.

12. Pamela Ferdinand, "Computerized Voting Eyed," *Boston Globe,* May 25, 1995, 40.

13. Theo Emery, "Upsets Seem Unlikely in Council Election," *Boston Globe,* November 5, 1997, B5.

14. According to Cambridge elections official Teresa Neighbor, rather than transfer partial votes strictly pro rata, the city uses an algorithm to identify specific ballots—1,499 of them in the Betsy Witherspoon example—for transfer to another candidate. A proportion is calculated by dividing the number of ballots cast for the candidate by the number of surplus ballots and rounding, up or down, to the nearest integer. In this example, 4,000 divided by 1,499 yields 2.668, which rounds to 3. Hence every third ballot for Betsy Witherspoon—the ballots were counted initially at random and numbered—would be transferred to the voter's second choice. Teresa Neighbor, e-mail correspondence, June 25, 1999.

15. Douglas J. Amy, *Real Choices/New Voices: The Case for Proportional Representation Elections in the United States* (New York: Columbia University Press, 1993), 10-11, 18-20.

16. See, for example, Robert Richie, "Full Representation: The Future of Proportional Election Systems," *National Civic Review* 87 (1998): 85-95, esp. 87-88.

17. Kathleen L. Barber, *Proportional Representation and Election Reform in Ohio* (Columbus: Ohio State University Press, 1995).

18. Ibid., 11.

19. Belle Zeller and Hugh A. Bone, "The Repeal of P.R. in New York City—Ten Years in Retrospect," *American Political Science Review* 42 (1948): 1127-48. In 1943 and 1945, when the United States was allied with the Soviet Union against Germany, voters had elected a pair of Communist Party candidates to the council. By 1947 Russia and the United States were no longer friendly, and many Americans viewed domestic Communists as dangerous subversives. Also see Guinier, *Lift Every Voice,* 264-65; and Richie, "Full Representation," 92.

20. George Pillsbury, "P.R. and Voter Turnout: The Case of Cambridge, Massachusetts," *National Civic Review* 84 (1995): 164-66.

21. Fain, "P.R. Elections in Cambridge."

22. Richie, "Full Representation," 90-91.

23. Guinier, *Tyranny of the Majority,* 15-16.

24. Robert Richie, "Cumulative Voting: Benefits and Drawbacks," *National Civic Review* 83 (1994): 214-16.

25. Richie, "Full Representation," 86-87. For a concise comparison of limited voting and cumulative voting, see T. Alexander Aleinikoff and Samuel Issacharoff, "Race and Redistricting: Drawing Constitutional Lines after *Shaw v. Reno,*" *Michigan Law Review* 92 (1993): 588-651, esp. 627-28.

26. Richard H. Pildes and Kristen A. Donoghue, "Cumulative Voting in the United States," *University of Chicago Legal Forum* 26 (1995): 241-313, esp. 260-72.

27. As political scientist David Canon notes, "Cumulative voting is clearly not a panacea for minority representation." See David T. Canon, *Race, Redistricting,*

and Representation: The Unintended Consequences of Black Majority Districts (Chicago: University of Chicago Press, 1999), quotation on 261.

28. Pildes and Donoghue, "Cumulative Voting in the United States," 272-302.

29. Guinier, *Lift Every Voice,* 28.

30. Richard L. Engstrom, Delbert A. Taebel, and Richard L. Cole, "Cumulative Voting as a Remedy for Minority Vote Dilution: The Case of Alamogordo, New Mexico," *Journal of Law and Politics* 5 (1989): 469-97.

31. Richard L. Cole, Delbert A. Taebel, and Richard L. Engstrom, "Cumulative Voting in a Municipal Election: A Note on Voter Reactions and Electoral Consequences," *Western Political Quarterly* 43 (1990): 191-99.

32. Robert Brischetto, "Cumulative Voting as an Alternative to Districting: An Exit Survey of Sixteen Texas Communities," *National Civic Review* 84 (1995): 347-54.

33. Lani Guinier, "Who's Afraid of Lani Guinier?" *New York Times Magazine,* February 27, 1994, 40-44, 54-55, 66, quotation on 54-55.

34. My description of Worcester County's brush with cumulative voting simplifies a complex saga of court decisions and appeals. On January 7, 1994, Judge Young ordered the county to present a plan to remedy illegal vote dilution. *Cane v. Worcester County, Maryland,* 840 F. Supp. 1081 (D. Md. 1994). On April 4, after county officials were unable to produce a compromise acceptable to the plaintiffs, Young called for cumulative voting. *Cane v. Worcester County, Maryland,* 847 F. Supp. 369. (D. Md. 1994); and Dave Kaplan, "Controversial Way to Vote Ordered," *Congressional Quarterly Weekly Report* 52 (1994): 857. On August 1 a three-judge panel in federal appellate court suspended elections under cumulative voting and extended indefinitely the terms of the sitting commissioners. "Plan to Use Cumulative Voting Prompts Election Suspension," *Congressional Quarterly Weekly Report* 52 (1994): 2373. On September 16 the appellate court ruled that the district court had exceeded its authority by ordering cumulative voting and remanded the case to Judge Young. *Cane v. Worcester County, Maryland,* 35 F. 3d 921 (4th Cir. 1994); and "Cumulative Voting Plan Dealt a Legal Setback," *Congressional Quarterly Weekly Report* 52 (1994): 2720. On January 6, 1995, the judge ordered single-member districts for the primary election and cumulative voting for the general election. *Cane v. Worcester County, Maryland,* 874 F. Supp. 687 (D. Md. 1995). On February 21, Young denied the county's motion to have his order stayed, pending appeal. *Cane v. Worcester County, Maryland,* 874 F. Supp. 695 (D. Md. 1995). On June 16, the appellate court vacated his order to use cumulative voting in the general election. Lyle Denniston, "Redistrict Plan Ordered in Worcester," *Baltimore Morning Sun,* June 17, 1995, 1B. The appellate court's decision was based on procedural issues, not the legality of cumulative voting. The decision, which partly affirmed and partly vacated Judge Young's ruling, was noted but not published in the court reporter; see *Cane v. Worcester County, Md.,* 59 F. 3d 165 (1995).

35. "Governor Chides Worcester Officials," *Baltimore Morning Sun,* June 4, 1995, 5C.

36. For a vigorous argument favoring cumulative voting to settle vote-dilution

cases, see Edward Still and Pamela Karlan, "Cumulative Voting as a Remedy in Voting Rights Cases," *National Civil Review* 84 (1995): 337–46.

37. *Cane v. Worcester County, Maryland,* 847 F. Supp. 369, 373 (D. Md. 1994).

38. "'Contrived' Ruling Ends Cumulative Voting in Worcester County, Md.," Southern Regional Council Web site, at <http://www.src.w1.com/vrrsum1996_worchester_nf.htm> [*sic*] (June 5, 1999).

39. "Cumulative Voting vs. Gerrymander," *Baltimore Morning Sun,* April 18, 1994, 6A.

40. George F. Will, "Sympathy for Guinier," *Newsweek* 121 (June 14, 1993): 78.

41. Neil A. Lewis, "Clinton Faces Battle over a Civil Rights Nominee," *New York Times,* May 21, 1993, B9.

42. "Withdraw Guinier," *New Republic* 208 (June 14, 1993): 7.

43. "Proportional Representation Fails," *New York Times,* April 24, 1993, 22.

44. Figures 10.2 and 10.3 are based on Karen Taggart, "Electoral Systems Survey: How People Vote around the World," at <http://www.igc.apc.org/cvd/cvd_reports/1995/chp7/taggart.html> (June 18, 1999). Compiled for 1995, the listing includes concise definitions of the various electoral systems and related terms. Taggart's list was also published in the Center for Voting and Democracy's *Voting and Democracy Report: 1995,* a collection of articles and essays about proportional representation. For additional information on international diversity in electoral systems, see Gary W. Cox, *Making Votes Count: Strategic Coordination in the World's Electoral Systems* (Cambridge: Cambridge University Press, 1997); and David M. Farrell, *Comparing Electoral Systems* (Hemel Hempstead, Hertfordshire: Prentice Hall/Harvester Wheatsheaf, 1997).

45. Mortimer's plan was apparently published in late March 1994 as a press release by the Center for Voting and Democracy, which he helped found. For maps and discussion, see Peter Applebome, "Guinier Ideas, Once Seen as Odd, Now Get Serious Study," *New York Times,* April 3, 1994, Week in Review section, 5; and Dave Kaplan, "Alternative Election Methods: A Fix for a Besieged System?" *Congressional Quarterly Weekly Report* 52 (1994): 812-13. Mortimer lived in Durham, North Carolina; for his words on the proposal, see Lee Mortimer, "North Carolina Takes Close Look at Modified At-Large Voting," *National Civic Review* 83 (1994): 83-84. Another champion of multimember districts for North Carolina was former congressman and presidential candidate John Anderson; see John Anderson, "A Way to End 'Political Apartheid,'" *USA Today,* August 3, 1994, A9. For a similar proposal, to minimize redistricting controversy in Alabama, see Gerald R. Webster, "Playing a Game with Changing Rules: Geography, Politics and Redistricting in the 1990s," *Political Geography* 21 (2000): 141-61.

46. Geographer Richard Morrill, who favors multimember districts able to "capture meaningful communities rather than arbitrary collections of blocks for the execution of elections," warns of districts "so large and heterogeneous as to extend far beyond typical real geographical communities." See Richard Morrill, "Territory, Community, and Collective Representation," *Social Science Quarterly* 77 (1996): 3-5, quotations on 4 and 5.

47. William Raspberry, "Superdistricts in North Carolina," *Washington Post,* April 17, 1998, A23. According to two prominent political scientists, the three-district North Carolina plan is a Republican plan; see Samuel Issacharoff and Richard H. Pildes, "All for One," *New Republic* 215 (November 18, 1996): 10.

48. "House District Standards," *Congressional Quarterly Weekly Report* 25 (1967): 2472. The bill also exempted Hawaii and New Mexico, which had two representatives each, but only for the 1968 elections. Although the 1967 federal statute is an impediment, there are no constitutional obstacles to proportional representation or multimember districts. See Richard K. Scher, Jon L. Mills, and John J. Hotaling, *Voting Rights and Democracy: The Law and Politics of Districting* (Chicago: Nelson-Hall, 1997), 293-94.

49. At-large elections in the eighteen states would have affected 259 seats. "House District Standards."

50. Tory Mast, "History of Single Member Districts for Congress," on the Center for Voting and Democracy Web site, at <http://www.igc.apc.org/cvd/cvd_reports/monopoly/mast.html> (June 28, 1999).

51. Steven Hill, "McKinney's Gambit: Will Proportional Representation Bring down the House?" *Humanist* 56 (January–February 1996): 5-6; and John B. Anderson and Rob Richie, "Proportional Representation as a Means to End the Redistricting Wars," *Legal Times* 20 (February 23, 1998): S45-46. McKinney's proposal was H.R. 2545 in 1995 and H.R. 3068 in 1997. For her views on the bill, see Cynthia McKinney, "PR for Congress," *National Civic Review* 85 (winter–spring 1996): 47-48.

52. For the bill's text and a brief overview, see Dan Johnson-Weinberger, "The Illinois Option Act," at <http://www.prairienet.org/icpr/Federal/IOA.html> (June 28, 1999). For praise of Illinois's use of cumulative voting, see Guinier, *Lift Every Voice,* 266-71.

53. Introduced in the first session of the 106th Congress, Watt's bill was tagged H.R. 1173. I found its text on the Illinois Citizens for Proportional Representation Web site, at <http://www.prairienet.org/icpr/Federal/HR1173Text.html> (June 28, 1999).

54. Michael Lind, "A Radical Plan to Change American Politics," *Atlantic Monthly* 270 (August 1992): 73-83.

55. Anderson and Richie, "Proportional Representation as a Means," S46.

56. Proponents of proportional representation largely ignore weighted voting, perhaps because they don't want their proposals to sound more radical than necessary. For discussion of weighted voting, see John F. Banzhaf III, "Weighted Voting Doesn't Work: A Mathematical Analysis," *Rutgers Law Review* 19 (1965): 317-43; Bernard Grofman and Howard A. Scarrow, "Current Issues in Reapportionment," *Law and Policy Quarterly* 4 (1982): 435-74, esp. 462-64; and Robert W. Imrie, "The Impact of the Weighted Vote on Representation in Municipal Governing Bodies of New York State," *Annals of the New York Academy of Sciences* 219 (1973): 192-97. In essence, Banzhaf rejects it, Imrie finds it empowering, and Grofman and Scarrow consider it a promising curiosity as long as populations do not vary too widely.

57. How party affiliation affects a representative's clout, and the power of his or her constituents, became readily apparent in the state senatorial district where I live. Veteran state senator Nancy Larraine Hoffman, who as a Democrat had enormous difficulty getting her bills through committees in the Republican-controlled senate, was appointed chair of the Senate Agriculture Committee a month after she switched her affiliation to the majority party. See Erik Kriss, "Senator Switch Major Move," *Syracuse Post-Standard,* December 20, 1998, B3.

CHAPTER ELEVEN

1. Although a color-blind society is a worthy goal, as Paul Peterson points out, "some racial classification is necessary [even though] the distinction between some and too much cannot be drawn logically." Paul E. Peterson, "A Politically Correct Solution to Racial Classification," in *Classifying by Race,* ed. Paul E. Peterson (Princeton: Princeton University Press, 1995), 3-17, quotation on 16.

2. David T. Canon, *Race, Redistricting, and Representation: The Unintended Consequences of Black Majority Districts* (Chicago: University of Chicago Press, 1999), 201-42.

3. See, for example, Charles S. Bullock III, "Winners and Losers in the Latest Round of Redistricting," *Emory Law Journal* 44 (1995): 943-77; and David Epstein and Sharyn O'Halloran, "A Social Science Approach to Race, Redistricting, and Representation," *American Political Science Review* 93 (1999): 187-91. For an early prediction of Republican gains, see Kimball Brace, Bernard Grofman, and Lisa Handley, "Does Redistricting Aimed to Help Blacks Necessarily Help Republicans?" *Journal of Politics* 49 (1987): 169-85. Not all political scientists buy this argument; for an opposing view, see David Lublin, *The Paradox of Representation: Racial Gerrymandering and Minority Interests in Congress* (Princeton: Princeton University Press, 1997), 111-14; and David Lublin, "Racial Redistricting and African-American Representation: A Critique of 'Do Majority-Minority [*sic*] Districts Maximize Substantive Black Representation in Congress?'" *American Political Science Review* 93 (1999): 183-86.

4. Pamela S. Karlan, "Still Hazy after All These Years: Voting Rights in the Post-Shaw Era," *Cumberland Law Review* 26 (1996): 287-311, quotation on 311.

5. Micah Altman, "Districting Principles and Democratic Representation" (Ph.D. diss., California Institute of Technology, 1998), quotation on 339-40.

6. Quotation from Web site Micah Altman's Dissertation Abstract, at <http://data.fas.harvard.edu/maltman/disab.shtml> (October 28, 1998).

7. For a fascinating inventory of the islands along the Atlantic coast and an overview of their history and connections to the mainland, see David Laskin, *Eastern Islands: Accessible Islands of the East Coast* (New York: Facts on File, 1990).

8. As Micah Altman observes, "The violations of contiguity in the latest round of redistricting far exceeded traditional baselines." Altman, "Districting Principles and Democratic Representation," 338.

9. See Kenneth C. Martis, *The Historical Atlas of United States Congressional*

Districts, 1789-1983 (New York: Free Press, 1982), esp. 52-61. In New York State, for example, the Seventh District for the Third, Fourth, and Fifth Congresses linked Clinton and Rensselaer Counties; the Eleventh District for the Eighth, Ninth, and Tenth Congresses linked Clinton, Essex, and Saratoga Counties; and the Eighth District for the Eleventh and Twelfth Congresses linked Clinton, Essex, Franklin, and Saratoga Counties.

10. For further discussion of the diminished importance of proximity in political representation, see T. Alexander Aleinikoff and Samuel Issacharoff, "Race and Redistricting? Drawing Constitutional Lines after Shaw v. Reno," *Michigan Law Review* 92 (1993): 588-651, esp. 637; and Dana Milbank, "Virtual Politics," *New Republic* 221 (July 5, 1999): 22-27. Two political scholars have even proposed allowing voters to sign up with a district outside the one they live in, but in the same state; see Deanna Marquart and Winston Harrington, "Reapportionment Reconsidered," *Journal of Policy Analysis and Management* 9 (1990): 555-60. Obvious complications include compliance with the one person, one vote standard and, especially in large states, the delivery of constituent services.

11. For further discussion of communities of interest, see Jonathan I. Leib, "Communities of Interest and Minority Districting after Miller v. Johnson," *Political Geography Quarterly* 17 (1998): 683-700.

12. Bernard Grofman, "Would Vince Lombardi Have Been Right If He Had Said 'When It Comes to Redistricting, Race Isn't Everything, It's the Only Thing'?" *Cardozo Law Review* 14 (1993): 1237-76, quotation on 1262. Cognizability is similar to the concept of "geographical coherence." See Aleinikoff and Issacharoff, "Race and Redistricting?" 637; and Ripley Eagles Rand, "The Fancied Line: *Shaw v. Reno* and the Chimerical Racial Gerrymander," *North Carolina Law Review* 72 (1994): 725-58, esp. 755-58.

13. Ibid., 1261.

14. Ibid., 1263.

15. Samuel Issacharoff, "Supreme Court Destabilization of Single-Member Districts," *University of Chicago Legal Forum* 26 (1995): 205-39, quotation on 214.

16. Ibid., 232.

17. Fully consistent with the Constitution, multimember districts and proportional representation are far less controversial than a redrafting of our state boundaries, which would require a constitutional amendment and thus seems virtually impossible, given the sacrifices required. Buried in the geographic literature are two examples of how the states might be restructured. Economic geographer John Borchert's little-known map of first-, second-, and third-order trade centers suggests the wisdom of replacing the forty-eight conterminous states with twenty-three provinces, each centered on a regionally significant business center. See Ronald Abler, John S. Adams, and Peter Gould, *Spatial Organization: The Geographer's View of the World* (Englewood Cliffs, N.J.: Prentice-Hall, 1971), 375. Although Borchert did not call for reconfiguring the republic, political geographer G. Etzel Pearcy left no doubts about the intentions underlying the double-

page cartographic centerpiece of his book *A Thirty-eight State U.S.A.* (Fullerton, Calif.: Plycon, 1973), map on 22-23.

18. Robert T. Gray, "Back to the Drawing Board," *Nation's Business* 87 (February 1999): 48-49; and Gregory L. Giroux, "The Hidden Election: Day of the Mapmaker," *CQ Weekly* 58 (2000): 344-51, esp. 345. State politicians were well aware of these likelihoods as early as 1996, when the Census Bureau announced intercensal population estimates; see, for example, Joseph Berger, "New York to Lose More U.S. Seats, Census Analysts Say," *New York Times,* September 30, 1996, B1. California seemed almost certain to gain; see Todd S. Purdum, "A Resurgent California Finds All That Glitters Is Its Future," *New York Times,* September 3, 1997, A1. Because of continued out-migration from the Delta region, Mississippi's loss will surely test the nonretrogression standard of the Voting Rights Act; see Frederick G. Slabach, "Race, Redistricting and Retrogression in Mississippi after the 2000 Census," *Mississippi Law Journal* 68 (1998): 81-103; and Jason Watkins, "Mississippi's 2002 Congressional Reapportionment: Legislators Beware—Eliminating a Minority Influence District May Violate the Nonretrogression Principle of the Voting Rights Act," *Mississippi Law Journal* 69 (winter 1999): 885-924.

INDEX